A Sacred Place to Dwell

Professor Henryk Skolimowski currently holds the Chair of Ecological Philosophy at the University of Lodz, Poland – the first such Chair of its kind in the world. He is the author of numerous books and articles, including *Living Philosophy* and *Eco-Theology: Toward a Religion for Our Times*.

This book is dedicated to those who are searching for the spiritual light and are determined to find it themselves.

A Sacred Place to Dwell

LIVING WITH REVERENCE
UPON THE EARTH

Henryk Skolimowski

ELEMENT

Rockport, Massachusetts • Shaftesbury, Dorset
Brisbane, Queensland

© Henryk Skolimowski 1993

Published in Great Britain in 1993 by
Element Books Limited
Longmead, Shaftesbury, Dorset

Published in the USA in 1993 by
Element, Inc.
42 Broadway, Rockport, MA 01966

Published in Australia in 1993 by
Element Books Limited for
Jacaranda Wiley Limited
33 Park Road, Milton, Brisbane, 4064

Cover design by Max Fairbrother
Design by Roger Lightfoot
Typeset by ROM-Data, Falmouth, Cornwall
Printed and bound in Great Britain by
Redwood Books, Trowbridge, Wiltshire

British Library Cataloguing in Publication
data available

Library of Congress Cataloguing in Publication
data available

ISBN 1–85230–443–X

Contents

Preface

A Sacred Place to Dwell is an extension of my basic vision of the world as a sanctuary, expressed in my earlier book *Eco-Philosophy: Designing New Tactics For Living*. In the course of completing that book it became obvious that ecological thinking has profound consequences for our religious views and our conception of God, for as the human mind and thought evolve so do our images and concepts regarding the nature and reality of God. This led me to articulate four stages of evolutionary thinking (see Chapter 4), the fourth stage of which is characterized by an evolutionary understanding of God, and an evolutionary God is the most radical consequence of our understanding of evolution.

How this ecological/evolutionary thinking applies to religion was outlined in my booklet entitled *Eco-Theology: Toward a Religion for Our Times*, and in the pages of this present book I share with the reader my perception of where we stand currently in religious and spiritual terms, as well as presenting several new ideas and interpretations. In doing so I leave aside the arguments of traditional theology, for they are more often a path of complexity than a path of light.

As the 1980s unfolded it became increasingly clear to me (and to many others) that we were entering a new era of human history, the ecological epoch, in which an ecological perspective will dominate, acting as a light to guide our understanding and our actions. The term *ecological* (derived from the Greek *oikos*, meaning home) includes the spiritual dimension – for our *oikos*, this planet, is not only our physical home, it is our spiritual home as well – and within this evolving ecological perspective religion cannot afford to remain static. It must evolve, itself becoming an agent for change and a source of new light.

The precise form the religion of the future will take is obviously difficult to describe in detail, but at the Congress of Assisi in 1986, in an extraordinary demonstration of harmony, five major religions simultaneously declared themselves to be friends of the Earth, concluding that as the saving of the environment is a theological problem of the first magnitude, it is also a religious problem of the first magnitude.

It would thus seem that various religious traditions are beginning to acquire a distinctly green shade or to be transformed into a new, more universal religion based on ecological spirituality. This means more than simply understanding the interconnectedness of all life. Religion in the ecological epoch will need to be based on a sense of deep communion with all beings – through empathy, through the power of the heart, through our deepest intuition of the sacred pulse of life and the sacred nature of the cosmos.

A more universal religion will certainly incorporate some of the essential characteristics of the spiritualities of the East, and yet it will speak to us with a distinctly different voice – a voice which will be acutely aware of the frailty of our planet, of the frailty of other species, and of the frailty of life itself. It will thus address our problems, our dilemmas and our agonies in a manner that is uniquely suited to our time. It will also revere Gaia, the Earth – Gaia is the name given by the ancient Greeks to the Earth goddess – inspiring us to understand the cycles of life, including our own, with a much greater depth and subtlety so that we shall cease destroying the environment through ignorance. Such a religion will have a profound reverence for all living beings, beholding all creatures and creation as sacred, and its very ethos will recognize that our own rights and those of other beings are of equal importance.

It will be a religion in which heart and mind combine, in which intuition and reason act in unison. It will fill us with such awe for the magnificence of the universe, with such a deep understanding of what is really important in life, that the present state of affairs in which economism is the dominant philosophy and consumption the ruling eschatology will be swept away. The economic god that we ourselves have created will no longer hold us in its merciless, exploitative clutches.

Furthermore, it will be a religion of celebration and joy. Joy does not mean vulgar fun, and celebration does not mean conspicuous consumption. Hopefully we shall attain sufficient maturity to realize that elegant frugality is a precondition of inner beauty, and we shall so radiate with positive energy that those who are afflicted by the angel of death will be carried with us on the wings of love and compassion. We shall cultivate a form of consciousness which naturally breeds peace, and through this consciousness the war-like mind-set so characteristic of the twentieth century will be changed. It will be realized that a leap of faith is a form of reason.

If this sounds far too optimistic for the so-called realists of the world, then they must examine their mind-set and language. What I have outlined so far is the evolutionary choice which confronts us. We cannot afford *not* to make this choice. Now, if following the path of realism means the death of culture and the spiritual death of human individuals, then this so-called realism is the *Tao* of extinction. It is not true realism.

No species wants to perish. Life loves life, and the pessimistic philosophy of pseudo-realism is the kiss of death. Optimism, hope, and a sense of celebration of the cosmos are the harbingers of true realism and the only possible path to the future. In the pages that follow I shall endeavour to articulate the possible future shape of religion, offering the reader a number of new ideas and interpretations, all of which I see as being the interweaving threads of the same tapestry – the spiritual life renewing itself through our own effort. In articulating what is struggling to emerge we not only elucidate the present, we also help to bring to fruition the seed of the future, for articulation is a vehicle, a tool of becoming. It is part of the creation of new art, of new philosophy and of new religion.

Among the new ecological/evolutionary ideas and interpretations I wish to offer the reader – with due humility – are the following:

- a new conception of divinity (as diffused throughout all creation) – justified throughout the entire book;
- a new conception of responsibility (we are co-creating with the universe/evolution/God);
- a new concept of spirituality;
- a new concept of God;
- a new concept of ethics;
- an interpretation for our times of the vision of St Francis.

In matters of creative and spiritual renewal we should not worry too much about clinging to our traditional concepts of God, for our concepts of God can only serve to limit the reality of God. Rest assured, God will take care of himself – so long as we take care of this beautiful planet which has been so gloriously created by the divine forces of the cosmos.

In fact the unfolding of human consciousness may lead us to discover the essential incompleteness of earlier religions. When that happens we shall discover a new dimension to God's existence through which he will reveal himself to us in a new form, speaking to us with a new voice – something which cannot happen if we burden ourselves with static concepts. We shall discover that God expects a new kind of prayer from us, the ecological prayer – one of ecological healing and reconstruction. And perhaps we shall also discover that true ecology is God's major project for us in today's world – a universal project of our times, uniting people of all races and of all religions by empowering us to co-create with him in bringing about a sustainable, just, balanced and beautiful new world.

CHAPTER ONE

Ecological Spirituality and its Practical Consequences

An Evolutionary Concept of Spirituality

Spirituality is a sublime subject. The greatest minds of the human race have struggled with it and left behind an illuminating literature. Yet we need to reflect on the subject again, simply because we wish to demonstrate that we are spiritually alive. Our circumstances and problems are unprecedented and they require a new spiritual response, a new form of spirituality. Older spiritualities were created in response to different problems, within the context of a different world view, and in order to articulate different dimensions of the human condition.

Spirituality is an articulated essence of the human condition of a given time. This conception of spirituality enables us to cherish various forms of spirituality in various cultures and religions. For it proclaims that *many forms of spirituality are justified*, and none can be claimed to be superior to all others.

Spirituality as an articulated essence of the human condition of a given time means that spirituality is not accidental but *essential* to the human condition, is one of the *defining characteristics* of the human condition. Without it the human condition cannot be truly human. Take away spirituality from the make-up of the human being and you end up with something less than a human. On this point, all world religions and major spiritual traditions agree. So crucial is the presence of spirituality in our humanness that without it our human status itself is in question. This can be seen clearly in our own tragic times, eaten up as they are by the cancer of materialism. The withering of spirituality in many present societies, and in particular individuals, is tantamount to a withering of humanity in us.

Although essential to the human condition, spirituality does not manifest itself in the same mould all of the time. It presents itself differently in different cultures and in different epochs. That simply means that *the essence of the human condition changes and evolves*. Some linguistic purists, or those attached to absolute schemes of things, might argue that the essence, by its very definition, does not change. But this is the old-fashioned, Platonic conception of essence. In our times we have come to recognize that every thing evolves and changes. *As the universe evolves it changes its nature*. Ilya Prigogine maintains that the *nature* of the laws of nature changes – as science evolves. There is hardly any scientific law or theory that has survived as immutable. All our scientific claims are seen nowadays to be tentative conjectures.*

Are the scientific laws but a fiction? Do they lie – as some people have suggested? No, they do not lie. In the realm for which they were designed, they are valid. The scientific laws, however, explain *only* what the scientific world view assumes. Scientific laws are at the mercy of the cosmology by which they were generated. Scientific laws and theories are thus self-referential. They articulate what the mechanistic cosmology assumes.

The great scientific laws have been challenged one after another, and found to be just approximations. During the last thirty years, we have seen the dissolving of scientific laws – as they were conceived in classical physics. The New Physics and quantum physics have not come up with any new universal laws. Quantum physics does not give us any clear pictures of reality. It is but a mathematical convenience, a set of symbols which relates to reality but obliquely. The idea of absolute laws was born out of dogmatic schemes of things, out of the frameworks (usually religious) which assumed that we can know the ultimate nature of things. Yet this purportedly ultimate knowledge has proved wanting over and over again.

If there are some cosmic laws or absolute laws, we are too ignorant at present to know them. We need to be aware of that fact and proceed with humility. To presume that we know the underlying laws of the cosmos, or even of the physical universe, would lead

*See especially: Karl Popper, *Conjectures and Refutations*, 1963.

to the arrogance and hubris which we have seen all too often among scientists, who have told us many times that they have completely mapped out the whole universe; only to retract from their positions twenty years later.

Now, since we live in a volatile and changing universe, it would be hardly surprising if the human condition were not subject to change. In truth we didn't need to wait for recent discoveries of science to realize that *the glory of different cultures and the enduring beauty of different religions are exactly the results of different articulations of the human condition.* For what is a culture if not a specific articulation of the various potencies which are contained in human nature and in the human condition? Each culture (if it is distinctive enough) weaves a specific pattern out of the sensitivities contained in the human condition.* And what is a religion, any religion for that matter, if not an articulation of spiritual potencies which are contained in the human soul and thus in the human condition? Each major religion weaves a specific pattern out of our spiritual endowment and out of our spiritual propensities. Each major religion represents and perpetuates a specific form of spirituality.

It has been argued, on the other hand, that all major religions ultimately converge and that their spiritualities are mirror images of each other. This insight is probably true. But it can only be confirmed in our innermost intuition when we are individually united with this ultimate oneness for which there is no word. *Only in the silence of our unfathomable souls can we intuit the quintessential unity of all things spiritual.*

In the world of different cultures, religions and tongues it is different. For in this world, in different cultures and different epochs, human spirituality is articulated and expressed in different forms. The form or expression of human spirituality is not unimportant. Different forms of expression lead to different conceptions of the human, to different lifestyles, to different ultimate aims or goals (eschatologies).

Religious fanaticism has been real enough – both in the past and in present times. There are some religions that have proved historically to be more tolerant, such as Buddhism; while there are other religions that have proved historically less tolerant (we might even

*For further discussion of sensitivities as maker of minds and of cultures see: Henryk Skolimowski, *The Theatre of the Mind*, 1985.

say prone to fanaticism) such as the Abrahamical religions – Judaism, Christianity, Islam. So the *form* of articulation of our spirituality does matter. It matters a lot during our earthly life. It also matters in the way it affects other people.

Our discussion so far has shown that our conception of spirituality – as an articulated essence of the human condition of a given time – is very useful for understanding the differences among world religions and their respective spiritualities. For each of them, when it emerged and acquired a distinctive form, was a child of its times, responding with great sensitivity to the problems and dilemmas of its time – while articulating the human condition in a new way. In summary, various traditional spiritualities are but various ways of articulating the human condition in response to new and often unprecedented problems.

Human beings respond to new problems in new, or at least ingenious, ways. Often the response is a variant of the practical way, which is by and large superficial. Religions and spiritualities, on the other hand, respond in a non-trivial manner; they respond in depth. Hence the difference between a technological response and a spiritual response. And hence a tragedy of our present time. Scientific-technological civilization wants to solve all the problems on the surface, with superficial answers. Whether there is such a form of spirituality which could be called 'technological spirituality' is an open question. If there is, it must surely be a very superficial spirituality. No wonder that people who search in deeper layers of their existence cannot find deeper answers through the courtesy of this superficial spirituality.

Our conception of spirituality as an articulated essence of the human condition also enables us to recognize spirituality outside the bonds of recognized religions. The established religions represent only a sub-class of all spiritualities. Before religions emerged, as distinctive entities, the human condition articulated itself in pre-religious forms of spirituality. Among the tribal people in the remote past, as well as in the present times (witness Australian aborigines but also witness the Khasis in Shilong, Northeast India, and Native Americans in the USA) people's veneration for nature was a form of their religion. Their spirituality was not centred on deities but *diffused* through nature. This represented a different articulation of the human condition. This form of articulation, and the spirituality corresponding to it, was by no means inferior to the forms of spirituality represented by organized religion – although Christian missionaries have claimed otherwise.

When we take Native Americans (or American Indians), then it is clear that their form of spirituality – centred as it is on the worship of nature and on reverence of all life – is by no means inferior to the religion and spirituality of the New Americans, who came to convert the savages but who, in the process, have been systematically destroying the whole sub-continent and slaughtering the Native Americans. Such a wholesale slaughter does not speak well for the 'superior' religion in the name of which it was done.

Spirituality is not about what gods we praise and how piously we do it, but about how our life affects other human beings, and other beings in the universe, including natural habitats and Mother Earth herself.

Why Ecological Spirituality?

As we look at the bleak harvest of our pursuit of progress, as we contemplate the destroyed forests, the destroyed cultures, the destroyed individual lives – all in the name of material progress – we realize that one of the causes of our half-blind economic pursuits is a total lack of reverence – for the world, for life, for higher values. The secular mind is completely denuded, insensitive, gripped by the fear of death and obsessed by the elusive criteria of efficiency.

The spiritual reconstruction of our time must start from the very foundations. How do we infuse reverence into a world which is conceived as a mere machine? Simply by denying that it is just a machine. The conception of the world as a clockwork mechanism is actually a strange creation. It is a new arrival on the stage of history. It did not exist in any culture before the seventeenth century. As a matter of fact, no proof has ever been given that the world is a machine, and nothing else but a machine. From the seventeenth century on, however, some people in the western world *chose* to believe that the world is a machine. They assumed this hypothesis, then acted on it. And then coerced others to accept that the world is but a machine. The result has been the nightmarish world we are witnessing at the end of the twentieth century.

But we don't need to accept this old-fashioned and clearly nihilistic assumption. The assumption is nihilistic because it leads to the destruction of natural habitats and the destruction of the inner lives of human beings. Instead, we can choose another metaphor through which to view the world.

Within Eco-Philosophy, which I am representing here, the world is viewed as a *sanctuary*. This perspective immediately changes a multitude of things. When the world is seen as a sanctuary, your role in it is that of a guardian, a shepherd, a responsible priest who maintains the sanctuary. From imagining the world as a sanctuary it is only one step to accepting yourself as a sanctuary – which needs to be treated with responsibility, care and reverence.

The universe conceived as a sanctuary gives you the comfort of knowing that you live in a caring, spiritual place, that the universe has meaning and your life has meaning. To act in the world as if it were a sanctuary is to make it reverential and sacred; and is to make yourself elevated and meaningful.

What the universe becomes depends on you. Treat it like a machine and it becomes a machine. Treat it like a divine place and it becomes a divine place. Treat it indifferently and ruthlessly and it becomes an indifferent and ruthless place. Treat it with love and care and it becomes a loving and caring place. A friend once asked me: 'What are you going to do to make people accept the assumption that the world is a sanctuary? Are you going to force them, or coerce them in some other way?' 'I am not going to do any such thing', I responded and continued, 'the most we can do is to say to the other: Don't you see that it is much more *natural* to accept that the world is a sanctuary rather than to assume that it is a machine?' In our intuition we respond to the idea immediately. It is much more intuitive to accept the world as a sanctuary than to assume that it is a machine.

By assuming that the world is a sanctuary we are not inventing a totally new idea. Rather we are going back to ancient times when people felt that the world was a *Misterium Tremendes*, a sacred place to dwell in. It is in this sense that the idea of the world as a sanctuary feels good. It is intuitively received as natural and uplifting.

Thus the beginning of our spiritual reconstruction is a reverential treatment of the world and of ourselves. This is also the premise of a deeper understanding of ecology. To understand the glory of all creation is to behold it reverentially. To understand fully the intricacies of ecological interconnections is to treat ecological habitats reverentially. To understand the beauty and integrity of the earth is to bow to it reverentially.

Reverence therefore emerges as a deeper understanding of ecology, of the earth, and of ourselves. Reverence is a principle of *understanding* – of the beauty of the world. It is also a principle of appropriate *behaviour* in the universe conceived as a sanctuary. The great symphony of life is singing through you. Listen to the song of life. We celebrate the miracle of creation by beholding it reverentially. We behold it reverentially by celebrating it. Life is not a terminal cancer but a great stupendous paean to be sung in joy at the altar of creation. Such are the consequences of the world conceived as a sanctuary.

It clearly follows that an in-depth comprehension of ecology is a total identification with it, is empathy fused with reverence. This is an important message for our times, namely that the true work for ecology is not only through campaigns to save this or that threatened habitat (though this is important too) but also in creating an attitude of mind within which the ecological and the spiritual are one.

Ecology is about the well-being of bio-habitats existing outside of man. Yet probed with a sufficient depth, *ecology is about the state of the human soul*. The interaction of the human with nature will be harmonious if the human soul is harmonious. We are one with nature. We are one with all creation. But this oneness will only be actualized in a harmonious and symbiotic manner when we realize that the state of ecology depends on the state of our soul.

Let us go a step further. Let us remind ourselves that at one time in human history the ethical sense was not fully articulated. Then the moral laws and moral commandments were created by such law-givers as Moses, Salon, the Buddha. Morality has become crystallized. *Human beings have articulated the ethical aspect of their existence*. Morality became a *dimension* of human life. In fact it has become a defining characteristic of being human. Whoever lacks a moral dimension is deficient in his/her humanity.

A similar thing happened with aesthetics. Some 100,000 years ago our human ancestors did not have a sense of the aesthetic as we have developed it in recent history. Then the sense of beauty, the sense of the aesthetic became articulated and consciously cultivated. *Human existence thus acquired the aesthetic dimension.* It is of little importance what form the aesthetic sense takes. The point is rather that to be human is to appreciate beauty. The *aesthetic dimension* is an integral part of our humanness in us.

In brief, the ethical and the aesthetic are part of our human nature. They partly define us. They are at the basis of our judgements and actions. A similar thing is happening with regard to ecology. Ecology as a form of thinking, and as an underlying system of values, is now becoming a dimension of our existence. To be a sensitive being is to be ecologically sensitive. To understand ecology – not superficially but in depth – is to acquire a new dimension of existence.

Let us look a bit further into the ways in which it becomes manifestly clear that the ecological and the spiritual are one in our times. *To worship God in our times is to save the planet*. Religious devotion which does not include a service to the planet is idle, if not idolatrous. Other aspects of the worship of God are important, but saving the planet is of supreme importance. This is the primary spiritual work of our times.

Let us emphasize: *healing the planet and healing ourselves is spiritual work*. Unless we see it as spiritual work of first importance, our efforts to heal the planet will be lukewarm. Working on ourselves, to get our act together, is important, but it must not be separated from the Great Healing Act that we all must do – healing the planet, healing societies, healing the mind. What we witness in our present forlorn and fractured world is the poisoned mind at work which in turn poisons everything around it.

If we lose the environment, we lose God. It is as simple as that. The holocaust through the destruction of the environment will be slow. But it may be as deadly as one brought about by nuclear weapons. Praying to Absolute Deities endowed with divine attributes is well and good. But to save the integrity and beauty of creation, we need to be engaged in *ecological prayer*.

Ecological prayer is acting out the ecological dimension of our existence, is bringing all beings into the wheel of our well-wishing. It is the meditative attitude of our daily action which helps to heal on all levels of being and is thus a form of ethics in action, based on solidarity and on compassion, which brings about a larger harmony in the long run.

The unity of the ecological with the spiritual may be seen in yet another way. The sacred books of many religions were written in different tongues and they worship the divine in different ways. Yet taken together, they may be seen as one big Bible, or one stupendous Upanishad, or one enormous Koran. When the differences are ignored, there is a theme that is running through all these sacred texts. This theme is the worship of the beauty and integrity of the planet. The 'Bible of the Cosmos' has ecological roots. *In the great sacred books, the ecological message has been written in such a small print that only recently did we learn to decipher it.* Thus, it took us a while to recognize that Reverence for Life and devotion to the planet are a focal point of religious teaching. But finally we have recognized this message. Hence, we see the greening of all world religions. They all declare themselves to be the friends of the Earth. And it is high time too.

The Practical Aspects of Ecological Spirituality

We venerate old spiritualities. Yet we need to dwell in a spirituality adequate for our times. What does it mean to say: 'a spirituality for our times?' It means a spirituality which is a part of the living substance of our times, a spirituality which responds to our problems, which sheds a significant light on our inner lives, which is a practical guide in daily action, and which helps to heal. Spirituality must be an inner mirror in which we can reflect adequately the shapes of our souls. Such a spirituality must also be a vehicle which heals.

Traditional spiritualities are superb mirrors in which our inner lives can be reflected upon and moulded into desirable shapes. However, older spiritualities are ill-at-ease with regard to our present problems – of individuals suffering, of the stress of the technological lifestyle, of the mother earth crying in pain. To witness our problems and our agonies is not to hide in a contemplative niche – although at times we need to renew ourselves in such a niche – but to find a new spiritual response to the splintered world we have created.

Ecological spirituality is such a new response. In a Buddhist way it confronts the suffering of the world. But it does so while recognizing the omnipresence of stress as a new form of suffering, and being cognizant that the healing of the earth (and of ourselves) is the most important practical and spiritual work of our times.

Ecological spirituality has a distinctive practical edge. It continually asks the following questions: How do you work on yourself so that you heal within? So that you heal your mind? So that you become a source of health and support for others? How do you help to heal the diseased educational system, the malfunctioning and carcinogenic social and political structures? How do you help to heal the earth in your immediate vicinity? How does your lifestyle help to maintain the equilibrium in distant parts of the world by not contributing to frivolous over-consumption?

All these questions are practical questions and spiritual questions at the same time. They require a new kind of awareness, a new kind of commitment, and a new kind of yoga.

Let us look at some of these questions a little deeper, particularly in the context of modern life. Let us see clearly how many more stresses are inflicted on us and how much more garbage is thrown at us in comparison with earlier, more simple and less stressful times. Let us contemplate how mercilessly our mind is invaded and twisted in many directions, and what we can do about it all.

Of all the gifts of evolution, mind is most precious. Yet we have allowed it to be manipulated by the concerns and interests that have nothing to do with our spiritual life. Look at ourselves. Look how much pollution is poured into our minds daily! Look how much trivia we have allowed to invade our minds. This garbage has trivialized our existence, is the cause of anxieties, is the cause of confusion that does not allow us to think right and act appropriately. If we care for the gifts of our life, if we care for the gift of our mind – *then we don't allow our mind to become a garbage container*.

The first principle of the right ecology of mind is: *don't allow your mind to become a garbage disposal unit*.

Conserving water is good. Not cutting trees is better. Eliminating air pollution is better still. But these are all half-measures. We have to change our mind which at present is rapacious and uncaring into a mind which is compassionate and caring. Only then can we hope to bring back blossoming to the Earth and radiance to our lives.

I am not saying that working to improve our environmental conditions is not important. I am not saying that working on the legislature to pass right bills is not important. These things are terribly important. But not enough in themselves.

To make peace with the planet Earth we must change the present mind-set, we must change the dominant consciousness which is greedy, parasitic and materialistic into caring, compassionate and participatory consciousness.

The second principle of the ecology of mind is: *No saving the Earth without changing our consciousness.*

What are you doing to save yourself from mental pollution? What are you doing to avoid mental noise, spiritual squalor, the conditions in which your life is debased? These are practical questions. These are at the same time spiritual questions.

The third principle of the ecology of mind is: *Be mindful of what you are doing to save your inner self.* This involves reflecting for five minutes every day at least on who you are, what is your highest potential, what kind of world you would like to live in. After you have reflected you must ask yourself how to act to bring about the kind of person you would like to be, to bring about the kind of world you would like to live in.

The ecology of mind leads you to acquire ecological consciousness,* which in turn leads you to right values, which are the foundations of right modes of action. We shall look at these ideas later.

The ideas we have discussed here are a part of the emerging *Ecological Philosophy.* What is ecological philosophy? Ecological philosophy is to be friendly with Nature and *knowing why*; is to be friendly with other forms of creation and *knowing why*; is to be friendly with your own body and *knowing why*; is to be friendly with your own mind and *knowing why*; is to be friendly with your inner god and *knowing why*; is to be friendly with the Great Call of Your Destiny and *knowing why*.

Ecological Philosophy means declaring and maintaining peace with yourself and with all creation. Unless this is accomplished we shall not save the earth. Saving the earth is a spiritual crusade of our times. *Ecology is a new metaphor for cleansing ourselves of all mental pollution.* While traditional religions talk about cleansing of the soul stained by sins, we talk about cleansing the mind stained by mental pollution. Our stresses are different, our Karma now takes different forms, our catharsis takes different forms as well.

This is the message of the *living* spirituality of the day. This message may seem different from the message of spirituality of traditional religions which calls for a contemplative attitude and a pious prayer. In our times God wants us to help to save his creation by saving the earth. In our own time the passive prayer must become

---

*For further discussion of ecological consciousness, see my book: *Living Philosophy, Eco-Philosophy as a Tree of Life*. See also chapter 2 of this book.

an active one, an ecological one. When we reflect on the lives of the Illustrious Ones, such as Jesus and Buddha, and in our times such as Gandhi and Mother Teresa, we see that their lives were an active prayer of healing, of helping, of nourishing and of nurturing. This is what living spirituality is all about, has always been, and will always be.

The Intrinsic Aspects of Ecological Spirituality

Although we have emphasized the *living* and *practical* aspects of ecological spirituality, we need to pay proper homage to its contemplative and intrinsic aspects as well. Each form of spirituality is the realization of our inner being, of our deepest potential, of the god within. The idea of spirituality as the realization of our inner divinity does not clash with the conception proclaiming that spirituality is an articulated essence of the human condition of a given time. For in different times the realization of our inner divinity may take different forms; this realization (the specific path to it or a specific form of it) may differ from individual to individual. Spirituality is as vast as an ocean and as diverse as the human condition itself. There are many roads leading to Rome, many roads leading to the inner god, many roads which actualize our spiritual potential.

Spirituality is about what we can potentially become. What we can potentially become is an aspect of spirituality conceived as the realization of our inner divinity. Paul Tillich and other theologians have emphasized that what a man is, is not a simple subject which can be described by empirical sciences. Indeed, describing a man as a brutish animal, only concerned with its physical survival, is a far cry from what the human is. The question, 'What is man?' properly understood should read as, 'What can the human potentially become?' With regard to this question, physical and biological sciences have little to offer. The question can be properly addressed in the realm of philosophy, theology, religion ... and poetry. Thus the poet Göethe writes:

> To treat man as he is
> Is to debase him.
> To treat man as he ought to be
> Is to engrace him.

What the human can become is a wonderful question. It is related to our spiritual potential but also to our will to make something of ourselves in our evolutionary journey. All great religions and enduring forms of spirituality are eloquent manifestations of man's will to make something more of himself than an ordinary bread-eater.

Now whether in his spiritual strivings the human has had an outside help is beside the point. The great spiritual advances were made precisely by those who helped themselves through an enormous inner discipline and through a stupendous act of will power – to transcend, to go beyond, to reach heaven.

This act of the will to transcend, this inner discipline to tune one's soul to receive the subtle melodies of the universe may be considered as an act of god-making, an act of actualizing the inner god. However, many of the illustrious ones speak of an intervention of the outside God which, as it were, begins to dwell in them. Yet with an amazing consistency those who were seemingly "chosen by God" are the very ones who have worked on themselves incredibly hard, and who have exhibited an enormous will power to actualize their inner potential.

Why do we strive spiritually at all? Why do we attempt to realize our inner divinity – when by not doing it, life would be much easier and more comfortable? The answer to both questions is *evolution*. It is in the nature of evolving life that it wants to make more of itself. It is in the nature of the evolving universe that it wants to make more of itself. It is therefore in the nature of human life that it wants to make more of itself. If life did not have this propensity to transcend, it might have got stuck on the level of the amoeba. If the universe did not have this striving to make more of itself, life might never have emerged.

Transcendence is therefore the formative force of the universe. Transcendence is the longing of the universe to make something of itself, and the will of the universe to do so. Transcendence is the primary expression of God's creative will. Transcendence (as the vehicle of all creative change) may be conceived as God itself.

The idea of transcendence is so powerful and luminous that it enables us to understand the movement of life in the universe as well as the vicissitudes of human spirituality. We humans are restless creatures. We always try to reach for new horizons. This thrust to go beyond is not an expression of our nervousness and schizophrenia (although sometimes it is) but an expression of the force of transcendence with which we are endowed as particles of the self-transcending universe.

Spiritual life is the blossoming of the force of transcendence. Beyond the biological and the mental, life needed further horizons. Life itself has been a formidable force compelling humans to go to the spiritual realm. Longing for immortality is an expression of transcendence, which tries to reach the infinite.

Human spirituality appears at this juncture of life when life breaks loose of the biological bondage and shines through with new qualities: the aesthetic, the ethical, the reverential, the spiritual. Spirituality as the flowering of life makes all life a divine phenomenon. To accept this conception of life is to accept the *evolutionary conception of divinity*. Yes, life had to reach out to the realm of the spiritual because it had the divine potency in it. *Within this perspective, God can be seen as a crystallized essence of the process of transcendence – finally leading to divinity.*

"We come from ashes and we return to ashes."

"We come from the Brahman and we long to return to the Brahman."

Neither of the two expressions is more justified than the other; and neither is true. Rather we should say: "We come from the seed and we want to blossom as all the seeds do." Blossoming is a visual form of transcendence in action.

Our discourse has shown that Transcendence, Evolution, God and Life are endowed with divine attributes; moreover, they are aspects of each other, they define each other, are reflections of the overall divinity of the universe.

The Indian poet Rabindranath Tagore was aware of the conception of life as divine and encompassing it all, as he sung beautifully in one of his songs:

The same stream of life that runs through my veins night and day runs through the world and dances in rhythmic measures.

It is the same life that shoots in joy through the dust of the earth in numberless blades of grass and breaks into tumultuous waves of leaves and flowers.

It is the same life that is rocked in the ocean-cradle of birth and of death, in ebb and in flow.

I feel my limbs are made glorious by the touch of this world of life. And my pride is from the life-throb of ages dancing in my blood this moment.

Rabindranath Tagore, *Gitanjali, Song Offerings*, No. 69

The practical imperative of ecological spirituality: "Heal the earth; heal yourself!" is in perfect harmony with the conception of life as divine and continually transcending itself. We have the divine heritage thrust upon us as our responsibility. Because our task is so enormous and so urgent, we need new strategies, new forms of perception, new forms of thinking, and ultimately, as I have argued, a new spirituality.

Thus, ecological spirituality emerges out of the awareness of the awesome beauty of the process that brought about life; of the awareness of the beauty of the human condition which is a vessel sheltering divinity of the universe; of the awareness of our responsibility to help creation to maintain its radiance. To heal the planet means to safeguard the beauty and integrity of Transcendence, of God, of Evolution – all combined.

Summary

Spirituality is an articulated essence of the human condition of a given time. Spirituality does not manifest itself in the same mould all of the time. The essence of the human condition changes and evolves. And so does spirituality. As the universe evolves it changes its nature.

Each major religion weaves a specific pattern out of our spiritual endowment and out of our spiritual propensities. Each major religion represents and perpetuates a specific form of spirituality.

The universe conceived as a sanctuary gives you the comfort of knowing that you live in a caring, spiritual place, that the universe has meaning and your life has meaning. To act in the world as if it were a sanctuary is to make it reverential and sacred; and is to make yourself elevated and meaningful.

When ecology is viewed with reverential empathy, the ecological and the spiritual are one. To understand ecology is to acquire a new dimension of existence. Ecological prayer is acting out the ecological dimension of our existence.

Ecological spirituality has a distinctive practical edge. It continually asks the following questions: How do you work on yourself so that you heal within? How does your life style help to maintain the equilibrium in distant parts of the world by not contributing to frivolous over-consumption?

Of all the gifts of evolution, mind is the most precious. Don't allow your mind to become a garbage disposal unit.

Transcendence is the formative force of the universe. Transcendence is the longing of the universe to make something of itself, and the will of the universe to do so. Transcendence is the primary expression of God's creative will. *Spiritual life is the blossoming of the force of transcendence.*

To understand religious devotion is to recognize that all religions are forms of worship of the beauty and integrity of the planet. The greening of world religions in our times is a clear indication that the call of the crying earth is heard by the churches.

The waning of religious forms of spirituality does not absolve us from the responsibility of the healing of the Earth, and from actualizing our spiritual potential. In spite of the religious crisis of our times, and perhaps because of it, we must have the courage to meet – in each of us individually – not only the Jesus of Nazareth but the Messiah of the Cosmos.

CHAPTER TWO

Spirituality and Consciousness: A Necessary Connection

From Eco-Philosophy to Eco-Religion

A new form of spirituality requires much more than the creation of a few high sounding precepts or new commandments. It is tantamount to the creation of a new world view of which a given spirituality is an integral part. To conceive of a new world view is a metaphysical endeavour which challenges human imagination to the utmost.

In a world dominated by atomistic understanding, the mind is conditioned to think small, and within the confines of narrow disciplines. In such a world the imagination which is capable of producing new metaphysical designs atrophies. The twentieth century philosophy, particularly that of the analytical orientation, is an eloquent case in point. This philosophy has produced many brilliant and penetrating minds but has been barren in terms of creative imagination.

In the mid-1970s, when the Ecology Movement started to flounder, it became obvious to me that what was at stake was not building new windmills or alternative "groovy" communities, but creating a new world view. The mechanistic world view could not sustain civilization. It had to be replaced with something else – quite radically different from it.

I set myself to work on this task – as others, from whom I expected illuminating answers, did not come up with satisfactory ones. In 1981 I published the book *Eco-Philosophy, Designing New Tactics for Living*, which was the first sketch of a new philosophy in the ecological key. My intention was to create an outline of a new world view which would be respectful of nature and of all other beings in the universe – cooperative and symbiotic rather than destructive and exploitative, and which would, at the same time, be a guide to meaningful life – which would thus resurrect human dignity and human meaning from the degradation in which they had been plunged by the technological system. The original small book must have hit the mark as it has been translated into twelve languages. It was gradually expanded and articulated to become *Living Philosophy: Eco-Philosophy As a Tree of Life*. *Living Philosophy* presents a *system* of eco-philosophy, spanning from eco-cosmology to eco-consciousness, on the one hand, and from eco-ethics to an eco-philosophical interpretation of power, on the other hand.

Now in 1981 or 1982, just after I published *Eco-Philosophy*, it became quite clear to me that a new metaphysics (in the ecological key) was not enough. Preserving the earth is important. But we also need to stretch our arms up to heaven. For heaven is part of the earth, and the earth is part of heaven – in our new, unified, interconnected world view. Thus we need to reflect *de novo* on the meaning of religion, of God, of spirituality. A new comprehensive metaphysics must, of necessity, include God, and the relationship of humans to the transcendent, to the sacred, to the ultimate. Thus the idea of Eco-theology started to form itself in my mind, and started to press for articulation.

At that time, in the early 1980s, there was no body of discourse on the subject to which one could relate one's ideas. So one had to work on one's own, aided by one's imagination. My reflections on God and religion, from the ecological perspective, were published as a booklet *Eco-Theology, Toward a Religion For Our Times*. The present book is a continuation and articulation of the discourse of that booklet. It is strange how slowly ideas grow and mature. What is so obvious now, namely, that we need an ecological religion, was not so clear even ten years ago.

While working on the early version of *Eco-Theology* (and ever since) I have been fascinated and perplexed by how complex and intertwined are our ideas concerning God and religion. When we think of God, we so often think (if only implicitly) about the entire world. It thus follows that a new image of God, or a new conception of religion, requires a new way of thinking about the world.

Now our thinking is very peculiar. It presents itself as a mere tool, a way of handling the world or God or whatever. But our thinking is not so innocent, is not a mere tool. It is profoundly shaped by our conception of God and of the world. So, it is often virtually impossible to think about God (and the world) in a new way while using the old modes of thinking. For this thinking, in a subtle and pervasive way, wants us to think about God in old ways, forces us back into the old moulds.

It is therefore important to realize – while we attempt to think about God in a new way and especially while we attempt to lay out a new theology – that our thinking and language may be a problem. Thus our thinking must be used with great care and a touch of distance. For so often it is a servant of the old. Furthermore our entire consciousness is a problem too. For it has been moulded, shaped and conditioned to serve the old and to be the mirror of the old.

There should be no doubt that our present consciousness is mechanistic in nature, as its function is to mirror and support the mechanistic world view, by which it has been formed and conditioned. Now while I reflected on the God of the ecological epoch (and a new theology in its image), I realized that the problem was not only with the articulation of the attributes of ecological God, but with the consciousness of individual living humans, who, as it were, wanted to think about God, religion and their destiny in a new way, but were held back, at the same time, by their own old consciousness.

In the process of working on Eco-theology, I had to articulate the features of a new consciousness (ecological consciousness) which is a prerequisite for thinking in a holistic, interconnected way; and which is a prerequisite for thinking about God in an ecological way – without being continually hampered by the strictures of the old consciousness.

But there was yet another stumbling block, and a considerable one too. For our consciousness is hooked, in a sense married, to a certain conception of the world. It operates within a certain picture of the universe, accepts certain arrangements of the world, a certain ontological order of the world. So you cannot just change your consciousness without changing the underlying cosmology. For if you do that, there is going to be a schizophrenic split in your mind: your consciousness will drive you in one way, while your knowledge (derived from your cosmology) will drive you in another way. Thus I had to formulate an ecological cosmology, as an alternative to mechanistic cosmology, as the ontological framework within which ecological consciousness and ecological God could dwell comfortably, and could really feel at home.

In any case, eco-consciousness and eco-cosmology had to be formulated as part of the metaphysical reconstruction, as essential and integral elements of the ecological world view – if this world view was to be a serious and comprehensive challenge to the mechanistic world. A detailed outline and articulation of eco-cosmology and eco-consciousness are presented in my book, *Living Philosophy*.

Spirituality and God do not exist in a vacuum, but are always embedded in the context of culture, of cosmology, of consciousness. Ecological God and ecological spirituality require ecological consciousness and ecological cosmology as their background. Only then will they be rooted. Only then will they be rationally justified, and felt existentially and psychologically acceptable. Only then will they be truly integrated into the new comprehensive ecological *Weltanschauung*.

Ecological Consciousness as a Foundation for Ecological Spirituality

The forerunner of ecological consciousness was the Ecology Movement, on the one hand, and various schools of Humanistic Psychology, on the other. In their respective ways they were against the temper of the mechanistic age. Both have emphasized holism and the irreducibility of large complex wholes (ecological habitats and human persons) to their underpinning components. Both these movements were a challenge thrown to the rationality of the mechanistic system. Both movements professed a new type of holistic rationality.

Moreover, in a certain sense both these movements possessed a religious flavour. They offered not only new intellectual vistas but a form of liberation. This liberation, although not always explicit, was meant to give us freedom from the deterministic and mechanistic shackles. Holism, which both movements emphasized, was the first step to liberation. Subconsciously we have grappled our way towards a new religion.

What are the chief characteristics of ecological consciousness? We shall enumerate six such characteristics and contrast them with the respective ones of technological consciousness. We shall not claim that these six characteristics completely define the scope and nature of ecological consciousness. We have to simplify. To understand is to simplify.

ECOLOGICAL CONSCIOUSNESS		TECHNOLOGICAL CONSCIOUSNESS
reverential	vs.	manipulative, controlling
holistic	vs.	atomistic
qualitative	vs.	quantitative
spiritual	vs.	secular
evolutionary	vs.	mechanistic
participatory	vs.	alienating

A more appropriate form of expressing the nature of ecological consciousness would be through a mandala as each of its characteristics is feeding into each other and feeding on each other; co-defining each other.

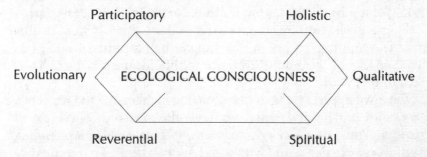

Mechanistic consciousness is not only descriptive – analysing and atomizing. In the very way it operates, it creates certain realities or, more precisely, renders reality in its own image. Moreover, mechanistic consciousness contains certain values, is a bearer of values. Let us look how it treats the world. Everything is reduced to mechanistic components, so that these components can be *manipulated and controlled efficiently*. The hidden underlying values are: control, manipulation and efficiency; and at a deeper level – a quest for power.

Conditioned by the imperatives of this consciousness, people, of necessity, have problems with love, with compassion, with embracing any significant form of spirituality. For love, compassion and spirituality are outside the realm of mechanistic consciousness. Those dimensions of human experience are simply not welcome in a universe which is governed by the criteria of ruthless efficiency, of cold rationality, of the desire to control and manipulate everything.

Therefore it should be absolutely clear that ecological conscious-ness must not aspire to be just another attempt to soften the hard edges of mechanistic consciousness. It must be an altogether differ-ent entity, an altogether different cast of mind. Against the quanti-tative, controlling and objective temper of the mechanistic mind, ecological consciousness proposes reverence as the chief *Modus Operandi* of the ecological mind.

Instead of seeing the world as disconnected atoms, ecological consciousness envisages it as one seamless web. As we perceive all things in the universe holistically, we also celebrate them as a part of the miracle of creation. To celebrate the world as a miracle of creation is something distinctive to ecological consciousness. This aspect is absent in mechanistic consciousness. Moreover, to cele-brate a miracle of creation is to behold the world reverentially. Thus we not only behold the world holistically; we also behold it rever-entially. First of all, by taking nothing for granted, by considering nothing as given, nothing as ours for the taking. Beholding all forms of life reverentially is honouring human beings and all beings as divine specks of creation. Reverence for life means the re-enchant-ment of the world.

When we appreciate truly the amazing alchemy of the universe, we cannot but be reverential vis à vis the awesome spectacle of creation. Thus reverence is an aspect of seeing and understanding the universe – in depth and with a true appreciation. To live in grace is to be in a continuous mode of reverential understanding. To live in grace is to think reverentially. To live in grace is to walk in beauty – as a Native American song proclaims.

Reverential understanding as well as a reverential attitude are not new creations of ecological consciousness. They have existed in traditional cultures and religions for a long time. We are only articulating them in a new way. The natural condition of the human person who is alive is to be enchanted by the world. Reverence is an acknowledgement of this enchantment.

When spelled out in detail and applied to all aspects of life and the world, ecological consciousness becomes a form of spirituality. For this consciousness is not only a form of perception but also a form of reverence; as well as a vehicle for the celebration of the universe.

Ecological consciousness is the rationality of ecological spiritual-ity. Ecological spirituality is the sanctum. The rational structure of the sanctum is ecological consciousness, which receives its benediction after it arrives at ecological spirituality.

Thus ecological consciousness is a vehicle (a precondition, a matrix) of ecological spirituality. Ecological spirituality is the ground for catharsis, for purification, for cleansing all that is polluting and damaging in ourselves and in our environments. When we think about all the healing that is needed on this planet, we realize that we cannot administer all the hundreds of cures that are necessary – as if they were separate cures for separate diseases. The one underlying cure for all or at least most of our maladies is ecological spirituality, as based on ecological consciousness and ecological values. Thus the application of ecological consciousness and ecological values in a serious and consistent way is tantamount to embarking on the voyage of ecological spirituality.

It is now clear why ecological consciousness is the foundation of ecological spirituality, for without it the latter is an abstract construct, not a living force capable of inspiring and guiding real people, whose emotions, hearts and minds are involved in their quest for meaning and spirituality. It is now also clear that certain forms of consciousness are conducive to life endowed with meaning and spirituality, and other forms of consciousness are not. Nihilistic, cynical or irreverent forms of consciousness, which are in the service of manipulation and exploitation, are not the right Tao to a life endowed with deeper meaning, and with spirituality. *For consciousness is a house for spirituality*. Beware of what kind of a house you have created, of what kind of house you have maintained, and of what kind of spirituality can find shelter there.

We have mainly discussed the reverential nature of ecological consciousness and the way it interconnects with and co-defines the holistic and qualitative aspects of eco-consciousness. Ecological consciousness is evolutionary in its nature. Why is the acceptance of creative evolution important to the overall structure of ecological consciousness and of ecological spirituality? And furthermore, why is the right comprehension of evolution important to our sense of the future, and of our destiny?

For three reasons. A sensitive reading of evolution informs us that the universe is quintessentially unfinished, that we are essentially unfinished. We have somewhere to go. We have a stupendous future in front of us. We are still toddlers in the cosmic playpen. We shall mature. We shall take our destiny more resolutely into our hands.

We shall become less stupid and more judicious, less vulgar and more sublime, less consumptive and more frugal.

Secondly, an intelligent reading of evolution informs us that evolution is a *divinizing* agent as it transforms matter into spirit. Some people claim that consciousness must have been present in matter and the entire universe from the beginning. Otherwise how could it have come to existence? I do not accept this view. If we take evolution seriously, we have every reason to believe that consciousness is *emergent*. It came to exist at a certain point in the development of matter.

Even if I could agree that consciousness could have been there – deeply hidden in the inner layers of matter waiting to be released – I would still argue that this process of releasing it from the bondage of matter was so extraordinary and so *creative* that we have every reason to say that this unveiling was a form of creation.

An intelligent reading of evolution is important for the third reason. It enables us to formulate the Middle Road, which lies in between religious consciousness on the one hand, and technological or materialist consciousness, on the other hand. The former claims that all divinity, and thus spirituality, is God-given and represents a reflection of God's divinity. The latter claims that spirituality and divinity are delusions or fictions of the human mind and that consciousness is merely a function of matter (Marxism). We claim that spirituality is an aspect of unfolding evolution. And that divinity is a flowering of evolution itself (we are back to the arguments of Chapter 1).

In proposing the Middle Road, ecological consciousness is not unlike Buddhism, which does not evoke any notion of God but which nevertheless assumes that we are spiritual and divine creatures, and that through our own work, through our own Karma, we can attain the levels of high spiritual enlightenment. This enlightenment comes through concerted practice, through the right tuning of the mind, through evolving one's own psychic capacities.

We are all possessors of our own divinity. But to release it from bondage, we need to perform a Herculean work of self-cleansing. Then we need to tune ourselves to the most evolved forms of human consciousness, as represented by the Illustrious Ones. Somehow in our busy lives, we tend to forget that all true progress, and especially spiritual progress, has been made by the *painstaking refinement of consciousness*.

In brief, creative evolution, as a component of ecological consciousness, is important for at least three reasons. It enables us to see that we are essentially unfinished, thus enabling us to make sense of the turbulent past, while looking forward to the immense promise of the future. It enables us to see that evolution is this process which makes matter into spirit, life into divine life. It enables us to follow the Middle Road of natural divinity which steers from the extremes and which recognizes us as both corporeal and spiritual, both rational and mystical – all within the bounds of natural evolution, which at the same time is a divine force.

How Do We Change Consciousness?

To change the human consciousness presently stuck in the mechanistic mores will require a painstaking effort. It will require some kind of revolution. This is how Ionesco defines revolution: "Revolution is a change of the state of consciousness". Incidentally, revolutions which did not work, including the Soviet Revolution, are the ones which failed to create a new consciousness.

Now let us consider another subtle but equally important point. Why is a change of consciousness so difficult on the individual level? Because the organism sees it as a challenge to its identity. We are comfortable in our old niches – whatever they are. Old niches are tantamount to stability. New consciousness implies dissolving the old structure, thus engendering instability to begin with, even if it leads to liberation and new freedom in the long run.

Working on ourselves is always a painful venture – even in spiritually oriented societies. It is particularly difficult in indulgent societies, such as ours. Yet, if we are going to make it, we need to change within. We are still postponing this act of cleaning up ourselves, which will be part of the act of cleansing the environment at large. Subconsciously we are waiting for a Messiah who will do it for us (see Chapter 6); or for some wonderful technology which will miraculously solve our problems. This is the heritage of messianic thinking and of technological thinking. The heritage of responsible thinking, on the other hand, tells us that we have to do it ourselves. And we shall – because we are intelligent and resourceful beings, particularly when pressed against the wall. And we are pressed against the wall! When we realize this fully, we shall go to our deeper resources. And we shall start reconstruction like the phoenix from the ashes.

How do we then change our consciousness?

1 • By realizing what are the pitfalls of present consciousness, why it misfires, what it does to our inner lives.
2 • By realizing what kind of consciousness is desirable and why; what the characteristics of the new consciousness are.
3 • By realizing what kind of methods and exercises may lead to the acquisition of a new consciousness.

Let us discuss these points at some length for they may hold the key to our individual inner reconstruction, and thus the key to our spiritual sanity, perhaps even the key to redemption. Francis Bacon discussed in his writings what he called "The idols of the market place" – common prejudices which cloud people's vision and right perceptions. He wanted to uncloud the human mind so that people could *see* for themselves. That Bacon's programme was too limited and that it ultimately failed is another matter. What concerns us here is Bacon's perception that those common prejudices of the epoch, which he called the idols of the market place, hold us in their powerful grip.

Look at the power of advertising and of the mass media! Don't you think that these media considerably and often surreptitiously influence and shape your mind, perceptions and values? Do these media want you to become a whole, connected and spiritual person? Far from it. So beware of the present idols of the market place – they are among the most powerful and insidious in history.

The idols of the market place are not the only forces that cloud our consciousness. Many spiritual traditions talk about the enormous capacity of human beings for self-deception. Deception is a shadowy aspect of human nature. It is caused by our self-interest, by our greed, by our lack of compassion – by and large, by *our smallness*. Deception is a form of blindness. *To see clearly* is to become a large person through cleaning the doors of our perception, but above all through purifying our conscience and enlarging our consciousness.

Ecological consciousness is simultaneously a vehicle for clearing the doors of our perception – obscured by the mire of technological decadence – and for enlarging our consciousness so that we regain the sense of our spiritual destiny, presently buried in the slothful orgy of consumption. Our larger spiritual destiny and our liberation from the shackles of consumptive consciousness are two sides of the same coin.

We have already started discussing point number 2 – outlining what kind of consciousness is desirable and why. Whether we call it ecological consciousness or not is not that important. What is important is to realize that technological consciousness is crushing and devouring us, and that we therefore need a new consciousness, radically spiritual in nature. *The re-birth of spirituality will not be possible without a re-birth of consciousness*. This is the most important message of the present chapter.

Now, a point or two should be added concerning the importance of the connection between spirituality and the *actual* modes of consciousness. Ex-communist countries of Eastern Europe, such as Poland, Hungary, Czechoslovakia, are now enjoying political freedom and religious freedom. They have vehemently rejected Marxist political doctrines, on the one hand, and Marxist atheism as the official religion, on the other hand. However, there are some problems with their rejection of atheism. It seems to be a verbal act only. They have not created a new consciousness which would fill the vacuum left by the Marxists. Poland is an especially telling case. It claims to be overwhelmingly Catholic. It prides itself on being the most Catholic country in the world – judging by the statistics of those who consider themselves Catholic. But it is all a facade. (I should know – they are my people.) *Their consciousness is not there*. Their consciousness is completely divorced from their avowed deep belief in Christian spirituality. Where their consciousness is, is an interesting question. It seems to be torn between their old allegiance to Marxist ideals of the fulfilment of humanity in material terms alone (which have been subtly but forcefully grafted on to their consciousness) and their new joy of re-discovering the "virtues" of rapacious capitalism. None of these two ideals leads to a path of any spiritual sanity.

Let us underscore the point. A negation of materialism does not necessarily lead to a new form of spirituality, let alone to a religious renaissance; especially if instead of a new consciousness we witness sliding back to questionable old-fashioned ideals, such as the belief in the Superior Moral Nature of the Free Enterprise System, or the belief in Thomist ethics, which proclaims the superiority of man over all other beings.

We should not suppose that rituals empty of content and void of any spiritual significance are especially characteristic of Christianity. Present Hinduism is another example of a culture in which high spiritual precepts constantly uttered are completely divorced from the actual modes of people's consciousness. The result is a perpetuation of words, prayers, devotional practices – while the actual mode of social interaction (particularly as seen in the actions of political institutions and the general corruption of social life) is so unholy. Judaism and Islam do not fare much better as far as the divorce of human consciousness from the living substance of their religion is concerned.

We have all been affected, people of all religions. TV and mass media have been brainwashing us all – by subtly imposing on us the parasitic, exploitative, manipulative, cold, detached consumptive consciousness. It is this consciousness that we need to transcend, if we are to witness a re-birth of religion.

Let us address point 3: what kind of methods and exercises may lead to the acquisition of a new consciousness? Specifically how do we practise reverence in a brutal, insensitive world? We need appropriate spiritual practices to make our consciousness reverential, whole, holistic and ultimately holy. In the religious traditions of the East many such exercises were devised and systematically practised. In the Christian tradition as well various spiritual exercises were devised, especially by Christian mystics. However our practices cannot be a repetition of some old, stale rituals, which have no bearing on our present lives. These practices must be vibrant with energy, appropriate for our present psyche, helping us to acquire a new peaceful reverential frame of mind in this rather turbulent, tense and disconnected time of ours.

While searching for such exercises, I have found many of the Buddhist practices useful and illuminating. There is something extraordinarily enduring even in the most rudimentary Buddhist meditations – perhaps especially in those ones. In every form of Buddhism, there is a meditation expressed in the words:

> May all beings be peaceful
> May all beings be happy
> May all beings be liberated.
> May I be peaceful
> May I be happy
> May I be liberated.

This is a simple but extraordinarily beautiful hymn of reverence, including all beings and ourselves. This simple hymn reverberates with universality and all-embracing compassion.

Now the acquisition of ecological consciousness and of ecological spirituality will require specific practices and exercises. To this end, my eco-philosophy has unexpectedly flowered in the form of *Eco-yoga*. Indeed I have felt myself compelled to create special spiritual exercises which are but an interiorization of the principles and ideas of Eco-philosophy. One of my friends, an actor and a director, has said, "if the ideas stay in the abstract part of your coconut, they do very little for you as a whole being". The translation of Eco-philosophy into *acts of being* is what Eco-yoga is about.

Any yoga is a system of exercises which aims at refining your consciousness and at bringing wholeness to your being. And so it is with Eco-yoga. Now the main point of a yoga is to practise it, not to talk about it. No description of a yoga is a substitute for the yoga itself. So practise we must. I practise Eco-yoga with groups of students and seekers in the remote mountainous village of Theologos on the islands of Thassos, in northern Greece.* There, in rugged but at the same time bucolic surroundings, amidst a scenery almost of Biblical description, it is easier to come to our inner selves and to see more clearly what are the conditions for a new spirituality and a new religion.

We need places of solitude, we need the inner silence in order to know who we are and what is our destiny. We need to detach ourselves from the machine (which is controlling our lives), escape from it (from its clutter, noise and pollution) at least temporarily, to be reassured that different modes of being and thinking are possible.

I am writing these words amidst the hills of the Central Himalayas. The huge, snow covered peaks are gracing me with their presence in the distance. They reassure me that majesty and beauty are aspects of my being. They inspire me to think high. They flood me with their radiant energy. They challenge me to become like them – unperturbed, lofty, enduring. They overwhelm me with their incredible size and grandeur. Yet they are also smiling benignly at

---

*For information about workshops on eco-yoga, write to: Eco-Philosophy Centre, 1002 Granger, Ann Arbor, MI 48104, USA

me saying: "All is well. All is well as long as you keep your gaze on
high horizons. All is well as long as you follow your destiny."

Summary

We cannot live with old myths. But we cannot live without myths.
We cannot live with the strictures and structures of old spiritualities.
But we cannot live without any spirituality. For life without any
spiritual foundation is a singularly arid and empty life – even if it is
a rational life. Our rationality cannot be a substitute for spirituality.
Rationality as the foundation of one's life leads to a barren and futile
life. For rationality is mute about the question of mystery, of human
destiny, of the manifestations of divinity.

When the human condition is inspired and led by the ideal of the
Buddha, the culture blossoms through Bodhisattvas. When the
human condition is inspired and led by the ideal of Christ – as it
was in the best periods of Christianity – then culture flourishes by
creating saints. In each case, the Buddhist saints and the Christian
saints were not created by the mere ideal of the Buddha or Jesus,
but by the pervading consciousness of the time, by the whole
context, including the world view and man's relationship to heaven
and earth, by which the human condition has been shaped and
determined.

When the human condition is inspired and led by the ideal of
efficiency and the control and subjugation of nature, then culture
expresses itself through countless mechanistic gadgets, which control the outside world, and ultimately control and regiment the
behaviour of human beings.

In order to change it all, we need to do more than curtail our
consumption, or humanize our technology, or develop science with
a human face, or become "kinder, gentler" to each other. These are
half-measures, at best. For our problems are spiritual and religious
– the problems of the soul led astray and starved. Thus we need a
new spirituality as a part of a new religion.

A new religion will not come *deus ex machina*, but will arise out
of a new consciousness. This consciousness itself will be an articulation of a new world view, which will be sympathetic to, and indeed
a gracious host to, a new spirituality and a new religion. Only then,
when we simultaneously re-define and creatively transform our

world view, our consciousness and our spirituality, will we be able to transcend the triviality and nihilism of our present condition.

We can create a new consciousness and we shall create a new form of consciousness, for this is an imperative of our times and an imperative of evolution, which does not want to become stuck in the consumptive mores of jaded and irresponsible people. With ecological consciousness as our foundation, we can articulate and justify ecological spirituality. For consciousness is a house of spirituality. Ecological spirituality is the sanctum. The rational structure of the sanctum is ecological consciousness.

CHAPTER THREE

New Values for a New Millennium

Ecological Values – Messengers of Eco-Spirituality

Since the Codex of Hammurabi, issued some 4000 years ago, human societies have tried to live according to some moral principles. These principles, above all, attempt to foster respect for other human beings, for life, for the divinity surrounding us. At different junctures of history and in different cultures, moral values are expressed differently. Yet the main focus is the same: moral principles are here to help us live with each other in harmony, to help us gain inner peace, to enable us to reach out beyond our petty concerns to transcendent realms beyond us.

Moral values express and articulate the moral ethos of a given culture, religion, or cosmology. They are the bridge connecting our ideals with day to day doings. They partake both in the transcendent and the immediate. They tell us what to do *now* in order to reach *beyond*, regardless of whether this beyond is transcendent God, the inner peace, or harmony among people.

In the past, religions often determined moral codes and were the guardians of moral values. Thus, within Christianity, the Ten Commandments dictated moral rules. In due time, the institutions of the Church appointed themselves as moral rulers. It has been a different story, however, during the last two centuries. The influence of religious values has waned while the influence of technological values has been in the ascendant.

What do we mean by technological values? Simply those values which follow from the mechanistic world view, from viewing the world as a large-scale machine. If you live in a world which you consider to be a machine and if you wish to be intelligent and purposeful in this world, what do you do? You try to understand the laws according to which the machine operates. Then you try to *manipulate* the machine *efficiently*. If you are brought up in an ethos which continuously attempts to inculcate in you the idea that if you are smart, intelligent and rational then you control the machine, then you are also bound to accept control, manipulation and efficiency as your chief values.

The world view based on the image of the machine, when pushed ruthlessly for a number of centuries, produces its own ethics of control, manipulation, domination, and cravings for power. These are distinctive technological values. These are the values which have come to dominate our consciousness in recent times. We exploit Nature and attempt to control and use it to our own advantage *because* it is part of the very ethics of the machine. The machine is here to control, manipulate, and generally exploit that which surrounds it. A manipulative culture produces manipulative and controlling, often power-crazy people.

All of us know, or should by now, that we as a human society cannot live by mechanistic values, for these values are antithetical to our freedom (yes, the machine does not care about your freedom), to our dignity, our inner peace, our quest for a spiritual destiny. Technological values simply undermine our very status as human beings.

Whoever still wants to argue that technology is good because it can produce some good things will argue very naively. The point is not what good things technology can produce *but what it has done to our psyche and to our values.* The point is that technological ethics, stemming from the image of the world as a machine, have created values which are destructive to ecological habitats, destructive to human societies, destructive to our inner lives. Whoever reasons with any consequence will be drawn to the inevitable conclusion that we must replace this whole ethos breeding insensitive, cold, uncaring, manipulative values – not just seek to humanize technology.

Religious values have waned, technological values are too ruthless to support us in our quest for meaning, dignity and beauty. Thus, ecological values enter the stage. Ecological values will be offered here as new intrinsic values, as values which can bind the whole human family together. Ecological values are not arbitrary. They are not yet another expression of fashionable relativism: if you are an ecologist, you believe in ecological values. If you are a hedonist, you hold hedonistic or consumerist values. Relativism is not a value-position. It is a sad denial of any value position. Relativism is sad because it confines the human to a limbo of non-values: if anything goes, nothing stays, nothing has any value. From relativism more often than not follow nihilism and cynicism – those cancerous values that eat away at our present society.

As I have already said, ecological values are not arbitrary *but follow from our reading of the imperatives of life at the present juncture of history.* When we look at our interconnected world perceptively, when we realize that we are all within the same web of life, then we cannot escape some simple conclusions. One of them is: You cannot foul your nest and not be affected by your own excrement. Put otherwise: "If you spit on the Earth you spit on yourself."

Whether your action is guided by your enlightened self-interests, or motivated by your care for the interests of future generations, the conclusion is so simple: it is stupid to undermine the foundations on which your house rests.

Thus some sort of ecological ethos follows from the very understanding of the situation in which we find ourselves at present. There are clear *imperatives* for action and for our behaviour which follow from our understanding of the constraints of life as lived in the world of limited resources. In the past, moral commandments were often given from above, from heaven, announced by God or some of his prophets. These commandments had to be accepted because God said so – regardless of whether your mind gave its assent to them.

It is not so with ecological ethics, whose main principles follow from below, from the bowels of life, of evolution, of the subtle web of the relationships that bind us together, and jointly nourish us. They follow from the very understanding of the beauty, interconnectedness and *frailty* of life.

Let me rephrase my point one more time, because it is so vitally important. We need to accept an ecological ethos not because it is imposed on us by someone who wants to make us good pious citizens, but because our understanding of the universe tells us so, because our reason tells us so. So if you are not obtuse, then you see clearly that ecological values are good strategies for living. I do not wish to chastise anybody or to coerce anybody into accepting ecological values, but only to emphasize how natural and reasonable they are for our present condition.

What are ecological values? How should we look at them? From our basic metaphor of the world as a sanctuary, immediately follows that the right and inevitable attitude towards the world is that of *reverence*. Thus reverence establishes itself as the chief ecological value. From reverence follow the values of: Responsibility, Frugality, Diversity, Justice.

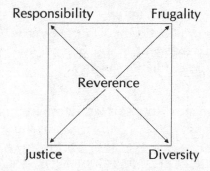

A true exercise of reverence immediately implies *responsibility*. Responsibility is one of the vehicles through which reverence is expressed and carried out. Responsibility should be seen as the wings of reverence. Responsibility is not "heavy", a burden, but a concept that gives us wings and thus enables us to practise our reverence as a cosmic dance. A more detailed discussion of responsibility will be carried out in Chapter 7.

Frugality follows both from responsibility and our sense of reverence. It is a relatively new ideal within the scope of intrinsic values. Therefore it needs to be discussed in some detail. Frugality is not to be advocated as an important ecological value to those who live in poverty and destitution, but among those who live in affluence. Affluence can create – and often does so inadvertently – a strain on others who may have been reduced into poverty through our overconsumption. In our interconnected world, and within its limited resources, if some consume too much, there is not enough for others. In a drastic form this idea was expressed in one of the Franciscan monasteries in California: "What you have and don't need is stolen from those who need it and don't have it."

Thus the ideal of frugality is born out of our care and responsibility for those others who are not invited to the feast of plenty. Frugality is born of our awareness that the orgy of consumption is obscene and immoral; of our awareness that in overconsuming we put an enormous stress on Mother Earth and therefore on ourselves in the long run. Frugality is born out of reverence for Mother Earth and for ourselves; out of our care of the integrity of the Earth, and the sustainability of life in the long run.

Frugality must not be confused with poverty, destitution or misery. Frugality is a positive virtue of our existence, chosen by those who understand it. Thus we can say simply:

- Frugality is grace without waste.
- Frugality is a pre-condition of inner beauty.
- Frugality is a majesty of simple means.
- Frugality is a joy of living simply.
- Frugality is a judicious and discriminate use of resources.

All the above are positive characterizations of frugality. They inform us that frugality *adds* to our lives, not subtracts from it; that great lives, such as Gandhi's and Mother Teresa's, have been beautifully frugal; that great works of art are – again – stunningly frugal in the simplicity of their means and the power of their expression; that reverence in our times requires stepping gently on the surface of the Mother Earth.

Frugality or voluntary simplicity is slowly becoming a way of life. More and more people are extolling its virtue in the mainstream periodicals. Some people are calling themselves with pride "the frugal zealots".

Diversity at first sight appears as an unlikely candidate for an important ecological value. Let us therefore look deeper at what it signifies. It signifies not only a description of things varied. It is thus not only a descriptive term. The machine loves simplicity and homogeneity. For this reason the machine wants to homogenize everything. Why? Because it then functions most easily and efficiently. Because we have homogenized (and impoverished!) the world so much, there is a danger – the danger of breaking the coherence and sustainability of life. Diversity is not only the spice of life. It is an imperative of vibrant life. *We must maintain diversity to maintain vibrant life.* Thus diversity, broadly conceived, is an imperative of life threatened by excessive homogenization, excessive reduction to mechanistic components. To emphasize diversity is a *moral* imperative, not only the necessity of preserving genetic varieties of crops, seeds and so on.

Evolution means diversity. Human cultures mean diversity. Fulfilling human lives means diversity. Diversity, as an elan of life, does not quarrel with frugality. It is through its evolutionary ascent that life has created wonderful diversity by means of dexterous frugality.

Justice has been enshrined in all significant value-systems of humanity. We know what justice is – particularly as it relates to us. We are fierce in demanding justice for ourselves. We are less adamant demanding justice for others.

Ecological justice signifies justice for *all* – not only within our own clan, and within our own society; not only among nations of people, but also with respect to all living beings; and with regard to the Mother Earth herself. Unless we bring all beings to the fold of justice, we cannot truly render justice to ourselves.

To render justice to ecological habitats and to far away people may seem at first a difficult act – particularly for those who govern themselves by individualistic and selfish ethics. But if we are in one web of life, if we take ecological spirituality seriously, then we come to realize that by rendering justice to all, we enshrine our own being.

Ecological values are offsprings of ecological consciousness. Ecological values are messengers of ecological spirituality. One needs to be aware of the intimate and continuous dialogue that is going on within the three realms: ecological spirituality, ecological values and ecological consciousness. On another level, one needs to be aware that an ecological God is using the three realms to bring a new sanity to the human race and a new radiance to the planet Earth.

In the remaining part of this chapter, I will extend the discussion of values in three different directions. I will try to show how important is the implementation of right values (especially that of justice) for global peace. Secondly I will discuss the importance of values for the emerging Green Politics. No political party, no political platform or forum can operate outside a sphere of values. Therefore we need to be aware what kind of values must be accepted to make Green Politics viable.

Finally, I will discuss the movement which has emerged around the idea of "Small is Beautiful", and will examine Schumacher's role in outlining a core of ecological values – to be practised within the compass of "small is beautiful". I will also address the question of why Schumacher was unsuccessful in creating and articulating an ecological theology.

Ecological Values as the Foundation for Peace

We are divided by different languages. We are divided by different ideologies. We are divided by our respective cultures which are often possessive and exclusive, and want to separate us from each other. Yet what we have in common far outweighs the divisions which we ourselves have created, sometimes inadvertently, sometimes deliberately. What we all have in common is the heritage of life, the planet Earth itself, the desire to live in peace and harmony and have a life endowed with meaning.

We are all aware of our common biological heritage, namely that all forms of life are built of the same building blocks, so that the life of a mosquito and the life of a lion (and to a lesser degree the life of a blade of grass) pulsate with the same rhythm of life. We are less aware of our *ecological* heritage, although of late an ecological consciousness has been gradually arising.

What is the difference between our biological heritage and our ecological heritage? The difference is subtle but important. The biological heritage accentuates the material aspects of life – the building blocks of life which are necessary for life to survive. Biology treats forms of life as energy machines. The ecological heritage, on the other hand, accentuates the conditions of the well-being of life, analyses the underlying matrix, the deeper structures which enable life to thrive and blossom.

The laws of biology are concerned with the survivability of particular individuals or particular species. The laws of ecology are concerned with the quality of life and with the maintenance of healthy *diversity* across various forms of life; are concerned with *optimal conditions* for various forms of life to live together.

The laws of biology are quantitative and expressed in chemical (or physical) terms. The laws of ecology are qualitative and expressed in teleological terms – the design of life and its purpose must be taken into account while studying the ecological heritage. Now let us unfold some of the hidden layers of the ecological heritage.

```
The heritage of  . . . . . . . . . . . . . . . LIFE
is the heritage of . . . . . . . . . . . . . . . INTERDEPENDENCE
The modes of interdependence are creative  . . SYMBIOSIS
The raison d'être for genuine symbiosis is . . . REVERENCE
```

Thus the very understanding of the complexity of life implies and necessitates the understanding not only of biological processes, but also of the deeper interconnecting structures which regulate and assure the well-being of larger habitats. In the final analysis we should understand that these deeper, interconnecting structures are laden with values.

Ecological values arise at this juncture of human history when understanding of life cannot be confined to the biological matrix only. Ecological values represent our understanding of those normative processes, within larger ecological habitats, which are responsible for the well-being of organisms; or, in more general terms, for optimal conditions of diverse ecological habitats.

Why should we care about many forms of life and not just only one, our own? Why should we care about symbiosis rather than allow one cancerous form of life to eat other forms of life? Because we are partial to the whole heritage of life! This partiality does not represent a scientific attitude but represents our value stand, our deepest commitment to the beauty and mystery of life. It should be emphasized that science, and its value-free descriptions of the world, *cannot* take any stand on the issue of values, on the importance of life, on the importance of the diversity of life. Diversity itself is an important concept. For (again) it is not only descriptive but also a normative one. We *value* symbiosis and diversity as vehicles assuring the vibrancy and resilience of life.

This analysis attempts to show that behind the idea of optimal conditions of ecological habitats there lies a set of ecological values which life has re-enacted over and again. My overall argument is simple, and is as follows: the underlying matrix of the ecological heritage of life (and the values embedded in it) are precisely the ones that can provide and assure the conditions of peace among people. *Ecology and peace are united* on this level of analysis when we understand the laws of the quality of life.

Obviously human societies are more complex than ecological habitats; at any rate they contain some layers of complexity which nature does not contain. I am not advocating a blind transplant of the laws and structures regulating the well-being of ecological habitats onto the present human world. I am rather maintaining that the implementation of the laws of the ecological heritage may be an important step to lasting peace. It is such an important step, in my opinion, that we *cannot* avoid taking it.

If we compare ecological values with the two other sets of values – *religious*, of pre-Renaissance Western culture, and *secular* or scientific values, of the present technological era – then we obtain the following picture:

Religious Values	Secular Values	Ecological Values
submission	mastery	reverence
worship	control	responsibility
grace	power over things	frugality
obedience	homogeneity	diversity
God's justice	individual justice	eco-justice for all

Let us underscore some main points. *Religious values are God-centred*. They regulate man's relationships to God; and to other human beings.

Scientific-technological values on the other hand are object-centred. *They simply regulate man's relationships with objects*. It is only dimly realized that scientific values have detached us from the human context and from the sacred universe. Let us emphasize it once more: the values which are most cherished in the advanced technological societies—manipulation, control, power, objectivity, atomization—have little to do with other human beings or with God.

Finally, *ecological values are universe-centred, and life-centred*. They reconnect us with all forms of life in the universe. They empower us and entrust us with responsibility for all. For we are a part of this grand sacred tapestry called cosmos. We are minute particles of this tapestry, yet terribly important as conscious weavers of this tapestry. Such is the message of ecological values.

Seen amidst the spectacle of chaos of our times, and amidst the indifference if not brutality of our behaviour, ecological values may seem too idealistic, particularly the value of reverence for life. Yet upon a deeper reflection we may have to come to the conclusion that it is precisely REVERENCE—for other people, for other cultures, for life at large—that may become the most important vehicle for establishing a universal concord for all living beings, for establishing peace on earth for all nations.

Amidst the force of chaos and disintegration, we cannot bring sanity and harmony by employing the same forces, but by seeking different strategies and forces. What unites us is the bond of solidarity, the understanding of compassion, the courage of reverence.

This analysis reveals further *why* ecological values can be seen as the foundation for peace: in enabling us to create a new social contract—which will be cooperative and symbiotic—ecological values pave the way to a lasting and just peace. In this sense, ecological values can become the spine of a new world order. And this will be an order based on solidarity and justice, not on hidden manipulation by Big Brother with a long stick. If so, then ecological values are of an importance second to none.

The heritage of life is immense and we need not apologize for learning from it, especially for learning from those structures and underlying grids of life which have assured its diversity and richness over billions of years. In promoting and articulating ecological values we are not inventing new fictitious philosophical entities but only unearthing the principles and structures which have proved life-enhancing in complex ecological habitats.

All life is a unity. We are a part of it. Since social life is a part of life in general, it must be governed by life-enhancing laws and principles. A new symbiotic social contract is imperative for a social life threatened by nuclear and environmental destruction.

Ecological values are trans-ideological, just as the oxygen we breathe. Ecological values may be viewed as part of a new unifying philosophy which we should wish to implement in order to survive.

Lasting peace among nations, and with all creation, does not mean the absence of war, or a temporary halt to hostilities. Peace as an overall harmony that enables all creatures to prosper is part of the agenda of all major religions, and surely part of God's design for creation to exist as God's creation.

The creative will of the universe requires peace as a condition for the continuation of evolution. Because in our times we have developed such powerful and terrible means of destruction, we need especially strong values to restrain ourselves from using our destructive potential. Furthermore, we need values that will direct us towards constructive agendas, towards the celebration of life. We need compelling values to lead us on to the road of harmony. Ecological values can then be seen as an expression of God's will to keep his creation intact, in balance, simply in existence.

The Spirituality of Green Politics

Why are Green Politics not more effective, although increasingly we see Green? Why are Green Politics in disarray? Why do the members of Green political groups quarrel constantly? Why are various Green groups unable to cooperate with others? Watching the Greens bickering with each other we witness a strange paradox. For on the one hand, the Greens and, generally speaking, ecology-inspired people proclaim and teach that things in nature are interrelated, in symbiosis, beautifully cooperative, woven into one web of life. On the other hand, in their actual behaviour towards each other, they are often antagonistic, belligerent, deserting their own ecological teaching about cooperativeness and sharing.

For strange historical reasons, many Greens have been attracted by anarchism. In the minds of many, anarchism and ecology are combined. In addition, some (if not many) have been inspired by the Marxist analysis of society, as they claim (quite rightly) that many of our social and political institutions are manipulated by rich elites at the expense of the middle class and especially the lower classes. Furthermore, many Greens, while rejecting the political right also reject religion and spirituality as they claim that religion has been used to the advantage of the manipulative privileged classes.

What emerges from this analysis is a composite picture of a typical would-be Green politician, who is a person of liberal persuasion, motivated deeper down by their adherence to some anarchist, Marxist, and anti-religious tenets. (Mind you, there is no such thing as a *typical* Green, for each of them is highly atypical, rebellious and at least slightly eccentric!)

Whether my composite picture is close to reality is not that important. The important point is that the Greens have taken anarchism much too seriously. Anarchism is not a political doctrine. Nothing can be built on the foundations of anarchism—nothing at all. Anarchism, as it appears in the historical perspective, is a futile game of immature minds.

I have seen anarchism applied semi-seriously in only one social reality. That was at Auroville in Southern India, an intentional community set up in the late 1960s. This community was inspired by the ideas of the Indian sage Sri Aurobindo. When asked about the future political system of humanity, Aurobindo replied: "It will be *Divine* Anarchy." After Aurobindo's death, the Mother (who was second-in-command) created an international community, Auroville, based on Aurobindo's spiritual ideals. Among those ideals was divine anarchy. *Divine* anarchy might have been able to carry the day. But ordinary anarchy is another matter. Divine anarchy means that each and every member of the community is fully mature, is spiritually connected and, in fact, enlightened. Among such souls, governments and traditional political institutions might be unnecessary as those individuals would be able to act in the manner that is favourable to all. We have not been able to create a society of such individuals. Therefore, the time for divine anarchy has not yet arrived.

In brief, the Greens have not yet reached maturity. By this I mean ecological maturity and political maturity. Ecological maturity means containing the individual egos for the sake of the larger whole. And my goodness, those individual egos flare up in the meetings of the Greens!

Political maturity means the realization that a true alternative to the present decaying political system lies *not* in abolishing all political structures, but in evolving new *spiritual structures* as the foundation of new politics. It is a condition *sine qua non* for Green politics to acquire a spiritual dimension. Green politics, to be a genuine alternative, must see in politics a spiritual pursuit.

The meaning of ecology derives from the Greek term *oikos*, home. Green politics and Green politicians must learn that our new *oikos* is not only a physical house, but a spiritual one as well. Furthermore, *oikos* must be viewed as taking the responsibility for our immediate house and for the whole planet, which is our enlarged home. We must treat this entire *oikos* as *Temenos*, which in Greek means a "sacred enclosure". Only when we develop a reverential attitude towards the entire *oikos* as *Temenos* will we be able to put Green politics on the right basis.

Political maturity for the Greens, and really for us all, means containing our egos, containing our individualistic rights, for the sake of our mutual cooperation, for the sake of the common good; really, for the sake of future generations and other species. Acquiring a new holistic vision and a reverential attitude is a necessary precondition of political maturity.

Maturity is a fruit of painstaking labour over oneself. In short, it requires a great deal of work on oneself. This is the inner work, ultimately the spiritual work. Whoever is not prepared to engage in this kind of work is unlikely to attain maturity. Working on our inner selves is terribly important in our times. We have mediocre politicians because *they are mediocre people*. We do not become people of substance and spiritual strength by embracing traditional manipulative politics, which ultimately makes us victims of the whole manipulative process. Many Greens have been sucked into the tube of manipulative politics.

Let us look at politics as a public domain. Politics is above all a public arena, in which various ideas and ideals struggle with each other. We might say, in a succinct way, that politics is the arena for the implementation of values. *Each and every political system is a carrier of values*. Each is inspired and guided by values, and each attempts to *inculcate* specific values on the populace at large.

Since Green politics clearly disapproves of, and publicly denounces, the values of the status-quo-oriented industrial societies, which are the values of unrestricted consumption and thoughtless exploitation of the earth and of other people, it must itself be *a vehicle of alternative values*, namely ecological values. Although Green politicians are aware that they must represent alternative values in their thought and action, somehow they are reluctant to talk about values explicitly, as if talking about values was bending backward to religion. *The shadow of secularism still clouds the thinking and axiology of Green politicians.*

Politics is an extension of eschatology. Green politicians must not be in the service of the eschatology of consumption, but in the service of the eschatology of self-renewal, based on our responsibility and co-creative partnership with the rest of the universe.

Politics is an articulation—in the public domain—of the prevailing concept of the human. Green politics must not endorse, even tacitly, the selfish, the egoistic, the greedy, and the consumptive image of the human. But it must strive consciously and relentlessly to project the image of the human person who is the responsible steward working diligently and compassionately in the world that is conceived as a sanctuary. The language of our discourse is important. If we shy away from the language which contains spiritual overtones and attempt to use the tough language of the present society and the present political system, inadvertently we bring with it the secular ideology and many of the values of the present status quo.

The human project is cooperative and spiritual. The politics of the twenty-first century, and of the centuries to come, must acknowledge and attempt to articulate the quintessential nature of the social contract, which cannot be social unless it is cooperative; and which cannot be human unless it is spiritual. Spirituality is an essential aspect of our humanity. Spirituality does not need to be confined to existing religions—as we have argued throughout this book.

Green politics must be in the forefront of this new understanding of the human project. We may therefore say that *Green politics is a spiritual crusade.* Until it becomes one, its message will not be convincing to ordinary people. People nowadays want a fundamental renewal, not the continuation of small thinking and of insignificant tinkering.

In brief, *Green politics must sort out its fundamental agenda concerning human destiny.* If it does not do so, it will remain insignificant wailing, full of sound and noble fury, signifying precious little.

Politics cannot be the handmaid of religion. For then it becomes an instrument of tyranny; as it has been in the past. Equally, politics cannot be the handmaid of selfish oligarchies, for then it becomes another form of tyranny; as we have witnessed in present times.

But politics inspired and guided by humane and spiritual ideals, while remaining in the relationship of symbiosis and universal sympathy to all beings of the cosmos, is certainly worthy of our aspirations and quests. Green politics can and should become an embodiment of the political will of the human race, an extension of the politics of evolution, and of life at large. So conceived, Green politics includes traditional democracy and at the same time transcends it.

The Spiritual Legacy of E. F. Schumacher

E. F. Schumacher (1909–1977) was an economist with an uncanny understanding of the deeper things that lie behind economics. He was a saintly person. His saintliness finally prevailed over his economic training. For a number of years, he was Economic Adviser of the National Coal Board in Britain—not exactly the position to prepare one to be a leader of people who challenge the existing civilization in an attempt to create a saner, simpler and healthier lifestyle.

His seminal book *Small is Beautiful* is deceptively simple. Perhaps we should remember that all great things are ultimately simple. So was Mahatma Gandhi's message. So is Mother Teresa's message. Schumacher should be placed among the spiritual lights of the twentieth century, with Gandhi, Schweitzer and Krishnamurti, rather than be viewed as an outstanding economist along the lines of Lord Keynes. In fact, Schumacher's system clearly transcends the Keynesian system as well as transcending much of traditional economic theory.

Those who see Schumacher mainly as an alternative economist or a promoter of soft technology miss an essential point—that his alternative economic system is embedded in values. Without understanding the underlying core of values, one is likely to fail to grasp the universality and attractiveness of Schumacher's ideas. Yet many have chosen to follow Schumacher on the economic/technological plane alone, insisting that if we reduce the scale and produce everything in the right scale (small is beautiful) and if we adopt soft, non-harmful technology, all will be well. Such an approach is superficial, for it limits itself to the physical while ignoring or neglecting the spiritual, the axiological (that which is related to values) and the eschatological (that which is related to the ultimate ends of human life).

Small is Beautiful is a beautiful example of ecological ethics in action. Schumacher does not spell out ecological values, nor does he even mention them explicitly, but his entire treatise is a poignant *application* of ecological values in various realms of human endeavour, particularly economics and technology. All the values we have distinguished as intrinsic ecological values, and as an element of ecological spirituality, are there in Schumacher's opus. It is therefore justifiable to claim that *Small is Beautiful* is a spiritual treatise. The fascinating aspect of this treatise is that the spirituality contained in it is expressed in such down-to-earth terms that it is almost invisible.

Yet when we look deeper, we cannot but be struck by the religious flavour of the whole book. What moves and attracts us, and makes the book universal, is Schumacher's passionate defence of the integrity and dignity of the individual, and of life at large. Thus reverence for life is the underlying foundation. In insisting that his book is one on economics "as if people mattered", Schumacher is adamant that we, the people, must take responsibility for our future, and not leave it to the machine.

Frugality is also enshrined throughout the whole treatise. The message is quite clear: gigantism leads to gigantic waste. Schumacher is particularly good in bringing forth the virtue of diversity against smothering and stultifying homogeneity—which is no friend to life resplendent.

Against the dictum of the throw-away society which revels in things plastic and then is overwhelmed by its mountains of garbage, Schumacher advocated the production of things which are durable, beautiful and simple (easy to operate). Durability is one of the keys to frugality and to preventing the waste which is overwhelming us. Let us be clearly aware that we are not only discussing things technological and economic, for the garbage that the throw-away society produces and then accepts as "a fact of life" spills over to our mental and spiritual life. The throw-away society wants us to accept pollution, including mental and spiritual pollution, as inevitable. While it has almost succeeded in suppressing older forms of spirituality, it has created its own sordid spirituality.

What is important to notice is the fact that Schumacher's overall message is so simple. It is back-to-the-basics: to realize that diversity means biological vitality; to realize that water comes from mountain streams and not from the tap; to realize that we should educate for life and joy, not for jobs; that repetitive jobs are killing us; that we must centre our lives and live within an appropriate scale. Although some of these recommendations sound like a down-to-earth practical guide to our complex living, there is a stream of values that underlies Schumacher's practical strategies.

Unlike any other economist in our times, Schumacher was convinced about the importance of beauty. Objects reproduced should be long lasting *and* beautiful. For beauty is life enhancing. Schumacher did not bother to provide any specific definition of beauty. He just insisted (like Plato did) that life is nourished by beautiful forms and atrophies in ugly surroundings. Schumacher's spiritual awareness has been his muse who helped him to sort out what is important from what is unimportant. (In Chapter 9, I will outline a new conception of beauty which shows *why* beauty is important to human life and to our spiritual quest.)

One of the most important of Schumacher's essays, in *Small is Beautiful*, is entitled "Buddhist Economics". This alone shows Schumacher's spiritual inclinations. 'When I finished this piece,' Schumacher confided to me a few months before his untimely death, 'I didn't know how to title it. First I entitled it "Common Sense Economics." But I decided against it because no one would take it seriously. Then I titled it "Christian Economics" and again decided against it, for everybody would think it was old hat. I finally decided to give it the title "Buddhist Economics", and everybody has loved it.'

Schumacher was in a sense destined to outline an Eco-theology for our times, as his thinking was ecologically and spiritually more mature than the thinking of those around him. He finally came to write his religious treatise *per se*. It was entitled *A Guide for the Perplexed*. Although deep in many ways, and shining through with wonderful insights, this book is ultimately disappointing.

The *Guide* is a restatement of Catholic theology, conceived and executed in the frame of reference of Thomas Aquinas. The treatise is disappointing, for many expected from Schumacher a religious statement for *our times*, uniquely expressing our dilemmas. Thus many had expected Schumacher to provide an outline of Eco–theology. With his supreme grasp of the interconnectedness of all forms of life within the ecosystem, with his exquisite sensitivities and deep religious convictions, he was almost pre-destined to be a religious spokesperson for the ecological era. This did not happen. *A Guide for the Perplexed* is an expression of the theology of yesteryear rather than the theology for our times.

We may wonder why Schumacher did not create an Eco–theology. The reasons are probably numerous. The most important is perhaps this: Schumacher was born a Protestant, yet throughout his adult life he was very attracted by Catholicism. As he put it to me, 'I have had a love affair with Catholicism for many years. Finally I decided to legitimize this affair and converted to Catholicism.' That happened in Schumacher's later life.

Thus the *Guide* suffers the symptoms of overzealousness, characteristic of new converts. We needed more than a restatement of Thomas Aquinas. Aquinas, though a stupendous thinker, is not really a very good guide for anybody who attempts to create ethics and spirituality for our times—which are ecologically sensitive —and who attempts to provide a new yardstick of justice which would mean justice for all creation.

During my conversation with Schumacher, I pointed out to him that many people would find his uncritical return to Aquinas a bit tiresome. "Maybe so," he said. "It is not important what form of religion you subscribe to. What is important is to return to our spiritual roots, to transcend present barbarism."

And yet the *form* of spirituality we embrace and live by *is* important. We clearly need a spirituality and religion for *our* times.

In brief, Schumacher's spiritual legacy is a mixed blessing. His explicit theological work, *A Guide for the Perplexed*, has not been influential at all, and will be of no historical significance. *Small is Beautiful*, on the other hand, has influenced millions, and will be exerting its influence in times to come, through its wisdom, through its courage, and its enduring values of ecological spirituality shining in action.

Summary

Moral values partake both in the transcendent and the immediate. They are the bridge connecting heaven with the Earth. They articulate the moral ethos of culture, religion and cosmology. As we seek a new culture, and attempt to embrace ecological cosmology and ecological consciousness as our new foundation, we are of necessity articulating new values, ecological values. The chief among these values are: reverence, responsibility, frugality, diversity, justice for all.

In the scope of new values, frugality, in particular, should be emphasized as it links our individual lifestyles, lived in daily reality, with the fate of the planet, and with the fate of many people in neglected parts of the world. Frugality is not an enforced poverty but a positive virtue. It is grace without waste. It is richness accomplished with slender means. Already Aristotle was aware of the beauty of frugality as he claimed that the rich are not only those who own much, but also those who need little.

As we are building a New Millennium, we must not forget the poor, the forlorn, the downtrodden. The quest for justice is a part of God's design. The quest for universal justice is an inherent part of our spirituality. All great spiritual traditions seek and promote justice. If they do not, then their spirituality is in question. Even if we are unable to bring about justice and eradicate the existing injustices, we must seek justice. This is our responsibility, our honour, our calling, and our mission.

Why do we link justice with ecology? Because injustice is a violation of peace. Justice, on the other hand, is an inherent part of ecological peace. To establish peace among all creatures of the planet is to establish universal justice.

Politics is the arena for the implementation of values. Each and every political system is a carrier of values. In order to bring back the battered Earth to blossom, we need a right political system, which will foster life-enhancing values. Green politics, as based on ecological values, is an imperative of our times. With the Greening of God, it cannot be otherwise in the realm of politics. Ecological Democracy, preserving and enhancing the rights of all, is a matter of common sense and historic necessity.

CHAPTER FOUR

Evolutionary God

The Four Stages of Evolution

To understand the meaning of evolution is the beginning of wisdom. We can look at evolution in so many ways. Whichever perspective we choose, one thing is certain – evolution has been nothing else but creative. This has been amply confirmed by recent discoveries in astrophysics and the New Physics. Our universe has been on a tremendous journey. This has been the journey of convulsive creation.

There is nothing static in our universe. Seen appropriately, the universe is one continuous story of extraordinary creative unfolding. And so is the story of evolution. The nature of our universe is awesome in its explosive transformations – from one fire ball to the billions of galaxies! From the first atoms of helium to the paintings of Rembrandt! What a story! What a spectacle of creation! Whoever does not see in evolution its wonderfully creative energies is on the wrong trail in understanding the world.

Yet we have a problem with evolution. It is *so* large. It cannot be contained in any definition. It is expressed in everything, but it cannot be expressed in words. In wanting to catch evolution in a net of words we are chasing the continually evasive phantom of becoming. How can we comprehend the totality of evolution, while we cannot express its meaning in crisp definitions? By pointing at this Enormous Phenomenon of Life in its various processes of becoming. The glory of evolution is the slimy little amoeba beginning to react to the environment semi-intelligently. The glory of evolution is the first eagle stretching its wings. The glory of evolution is the first monkey using a stick as a tool. The glory of evolution are the Vedic hymns conceived in silence and expressed in an ecstatic rapture. The glory of evolution is the monumental *Principia Mathematica Philosophia Naturalis* of Newton, attempting to express all visible nature in quantitative laws. The glory of evolution is our reflective mind reflecting on the glory of evolution.

A terminological note: when I speak of evolution, I do not merely mean the Darwinian kind of evolution; even less so do I mean social Darwinism (which is a form of ideology, supporting injustice and inequity). In short, I do not mean any deterministic process (which for example Jacques Monod postulates in his book *Chance and Necessity*) that makes us victims of iron necessity and/or capricious chance.

I use the term 'evolution' in the sense in which Bergson and Teilhard applied it, as a partly creative process which alone can account for emergent qualities and new forms of life. All *significant* changes in the history of the universe are the result of evolution so conceived; 'significant' invariably means within the compass of our understanding. Thus evolution is inextricably tied to our understanding of the development of the universe, and our understanding of our place in it.

Everything evolves. So does our thinking about evolution. It has evolved significantly since Charles Darwin came on the stage. As it was important in Darwin's times to see the unity of mankind with lower forms of life, so it is important in our times to see the differences between the human beings and lower forms of life. Evolution means differentiation; means growing complexity and growing consciousness. Whoever is blind to those characteristics of evolution is blind to the very *raison d'être* of evolution.

Let us therefore attempt to see how our thinking about evolution has evolved, and how it is evolving now. In so far as we are evolution conscious of itself, we have the responsibility to help this process of evolving.

To begin with, the discovery of evolution does not start with Darwin but with the geologist Charles Lyell. Lyell saw and described the geological evolution in his seminal treatise *Principles of Geology* (1830–33). By the time Darwin came onto the stage, the ground was prepared. People were ready to entertain ideas about evolution. Their imagination could *conceive* that the world was evolving. Darwin applied Lyell's idea a step further and showed that species were evolving as well. Thus we witness the first two stages in the discovery of evolution:

1. Geological (Lyell)
2. Biological (Darwin)

The next two stages of this discovery are happening under our very eyes. We are actually articulating them, sometimes consciously and sometimes only gropingly. These next two stages in our discovery of evolution are the recognition of *conceptual* evolution, and then of *theological* evolution (the latter, because of the nature of traditional religions, is most difficult for people to accept).

3. Conceptual
4. Theological

The recognition of conceptual evolution (3) is based on the realization that all our knowledge is evolving, that our mind and our knowledge are not fixed – once and for all. Put otherwise: there are no absolute laws of science which ultimately describe physical reality. All knowledge is conjectural (Karl Popper). The New Physics goes a step further: the known and the knower merge. As our minds evolve so will our knowledge, as well as our 'laws' of science. ('The nature of the laws of nature changes' – Prigogine.) The recognition of the fallibility of all human knowledge, including scientific knowledge, was just an extension of the evolutionary perspective into the realm of knowledge and of conceptual thinking.

The lion's share of credit in this respect goes to Karl Popper who, in his *Conjectures and Refutations*, tirelessly argued that fixed scientific laws are an illusion. We only have conjectures, tenuous and tentative, holding for a while, and then they are replaced by other conjectures. It should be noted that Popper himself did not perceive that his epistemology was an extension of the evolutionary theory into the products of our knowledge. This interpretation has become possible only with the advent of the New Physics. The New Physics exemplifies the third stage of our evolutionary thinking: evolutionary thinking applied to thinking itself; *seeing our minds as a part of the evolutionary process*.

The recognition of theological evolution or the discovery of evolution in the realm of our thinking about God (4) means that we no longer accept one fixed and immutable God. We thus recognize that our concept of deity, of redemption, of salvation are evolutionary products too. We shall need to spell out this insight in some detail. As evolution goes on, our being changes. As we change, our minds change. As our minds change, our deities change, our thinking about ultimate concerns and ultimate anchors changes.

We can re-trace the chronology of the four stages in the development of our evolutionary thinking. In 1830 evolutionary thinking is applied to geology (Lyell), in 1859 to biology (Darwin). Then fifty years later some thinkers (Poincaré, Duhem) become aware of the importance of language in shaping knowledge and indirectly in shaping our world. Another fifty years pass and in the early 1960s (with Karl Popper and Thomas Kuhn) we witness the articulation of the third stage of evolution – conceptual evolution. In 1985 Henryk Skolimowski articulates the fourth stage – evolutionary thinking as applied to religion and the reality of God; he also articulates explicitly the four stages.

Let us emphasize the main point. We have but the human mind, both for the understanding of human affairs and of celestial beings. Our mind limits the forms of our comprehension. As our mind comprehends so we understand. Our understanding of God leads to specific conceptions of God. These specific conceptions of God outline *the reality of God*. There is no other reality of God but one that our mind is capable of constructing through its powers. Our mind is the final terminus – not the mind understood as an abstract analytical brain but as the sum total of knowing and comprehending faculties. As our comprehension changes so change all realities – as apprehended by the mind. When the mind has discovered the evolutionary dimension of all there is in the universe, this is bound to affect its conception of knowledge and its idea of the reality of God (cognitive evolution and theological evolution respectively). To say it once more, Eco-theology (as well as Creation Theology) represents the fourth stage in the discovery of evolution.

The two latter stages in the discovery of evolution are so close to us that we do not have a sufficient perspective to see them clearly. The issues are not only new but also contentious to the point that many people, conditioned by traditional understanding, will raise their eyebrow more than once. Some, particularly those representing interests of established orthodoxies, might even want to argue as follows: "We know what the reality of God is. The views that depart from the established orthodoxies represent heresy." They might further respond: "There is a tradition to observe, the dogmas to follow." Yes, there are the dogmas. But what do they have to do with the reality of God? The reality of God is an immensely difficult subject. We are far from comprehending this reality. As we develop and mature we will understand it better.

Some Consequences of Theological Evolution

What are the ultimate consequences of the notion of evolution as continually emergent, as ceaselessly perfecting itself through its own effort? One of the consequences of taking evolution seriously is the realization that *there never was a paradise before*. We came from brute inauspicious beginnings. The primordial atoms and the slimy amoebas did not live in a paradise; nor did the fish, the monkeys, the hunter-gatherers.

We have made our spiritual ascent very gradually and painfully. Our present imperfections can be well understood in the light of our groping through evolution. This journey also explains our striving for perfection. This striving is the nature of evolution. *As it goes along, evolution wants to make more and more of itself.* Transcendence is the formative force of the universe. And so it is with us – we want to make more and more of ourselves. We aspire to divinity. We attempt to make ourselves divine. But the process is slow. The actual divinity comes at the end of the road, at the end of time. We become godly if we actualize the god within. Such is a truly evolutionary reading of our divinity.

God is spirituality actualizing itself through us. The idea of God within makes perfect sense. The further we go in our evolutionary journey, the closer we may approach him. Our journey therefore is to transcend – further and further, and never return. *A return represents a fall from grace. Thus evolution resolves the intractable dilemma of traditional religions: how to explain man's imperfections while at the same time claiming that he/she is a divine being.*

Is there a way of incorporating the Christo-genesis (so dear to Teilhard) and the Brahman-genesis (so dear to Aurobindo) within the consistent evolutionary scheme? Yes, there is, namely by treating Jesus not as God, and by treating Brahman not as the Absolute Ground of Being, but by treating both as *symbols*, as *inspirations*, as *reminders* that even at this early stage of our evolutionary journey, we are capable of much grace and divinity.

We invest our deities with the most illustrious attributes we desire to possess, and then through the emulation of these attributes we make something of ourselves – as human beings and as spiritual beings. *Our humanity is a product of mirroring in our lives the qualities we have vested in our deities*. The role of religion in the symbolic transformation of the human has been second to none. This is a salutary aspect of traditional religions. We can see from this discussion that the acquisition of spirituality and the rise of religions as tools of human perfectibility are thoroughly consistent with the evolutionary design. It seems so clear. We don't need a God at the beginning of the journey in order to recognize our journey as divine. For in truth, evolution is divine in its nature and in the countless miracles it has produced. *Evolution is God. God is evolution.* And who can deny that?

In the post-positivist age we cannot throw out our rationality although we do not want to be victimized by it. Rationality is one of the wonderful acquisitions of evolution. Thus in our times we wish to simultaneously preserve our rationality and our spirituality.

In the post-Newtonian age, when we realize that we co-create with the universe, we cannot relinquish our responsibility to technological experts or to Almighty God. The ultimate expression of our responsibility is to realize that *we have no choice but to accept the idea that we are God in the making*.

In the post-materialist age, when we can no longer accept the idea that God is dead, we cannot, at the same time, accept the old images and symbols of God as the Absolute Ground of Being. Therefore, we need to develop a spirituality and divinity appropriate for our times.

We are living in extraordinary times when it is possible to be rational and spiritual at the same time. This is indeed the imperative of our times. This is also the result of a new reading of evolution, namely that the rational and the divine are not antithetical to each other but aspects of each other, that is, within the universe of creative evolution. In previous epochs we did not have this luxury and therefore we were torn between the two. Now the time for harmony and healing has arrived. The right reading of evolution is a healing process. For the right reading of evolution is participating in God-making.

Why Are We Afraid of an Evolutionary God?

God is not afraid of anything. God is not afraid of his evolutionary nature. Only we are.

In the world that was static and stagnant, we thought that perfection meant immutability and the state of frozen being. In the world exploding with creative energies and unexpected vistas, our conception of perfection of God needs to be re-examined. If God is creative, if God possesses all the creative potential we can conceive of, why would not God be one which is evolving?

Even if we apply common sense, it stands to reason that the being which is continually creative and unfolding is more perfect than the one which is static and frozen. Indeed why should we deny to God *the joy of creative becoming*? Why should we not assume that God *delights* in his powers and abilities of self-transformation? By merely saying that God is not the kind of being that changes? But what do we *really* know about the nature of God? God is everything. Therefore God is also part of creative becoming.

Why are we afraid of evolutionary God? There are at least five reasons. Even if these reasons were somehow justified in the past, they are no longer justified in the present times.

The first reason has to do with our inertia and laziness. We have got used to a certain conception of God. We are comfortable with it. We prefer repeating old dogmas to re-thinking our condition. We are lazy at heart and prefer easy solutions. Re-thinking the nature of God is a difficult yet a magnificent challenge at the same time. It is a magnificent challenge because it affects so many realms of our being. It is such a difficult challenge for precisely the same reason. If and when the process of re-thinking the nature of God is completed, we are no longer in the same universe. For this reason the challenge is awesome. By and large, we are too indolent and too lazy and we do not gladly undertake a fundamental re-thinking of any significant problem.

The second reason is the power of tradition. Tradition, as we know, is a force to reckon with. It often breaks the neck of new ideas and new departures. But finally the tradition yields, is broken by the tide of the new.

The Christian tradition (and equally the Hindu tradition) is centred around the conception of God as immutable, as an absolute ground of being. This tradition has been significantly reinforced by the powerful Church which has been the watchdog of the purity of the creed. To understand how powerful the Church has been, it suffices to remember one instance – the trial and burning of Giordano Bruno in 1600.

Now, given the overall power of the Christian tradition, and given the intimidating nature of Christian theology, it would seem a daunting task for anyone to propose a new conception of God and not be annihilated by orthodoxy; or at least shouted down by numerous critics. The striking fact in our times is that orthodoxy is no longer sure what to believe in. Besides, the conception of evolutionary God may not be as heretical as it may at first appear – if we think through the nature of God, which we must.

The power of tradition is great indeed. But it is not nearly as great as tradition itself assumes. For in time, it will inevitably be broken, altered, changed. *All traditions are evolving!* If a tradition wants to survive, it must evolve. The same holds for orthodoxy. As new interpretations emerge, orthodoxy is changed. The history of the Christian Church has been the history of heresies. These "heresies" have been incorporated into an orthodoxy which has been continually changing, thus *evolving*.

The third reason why we are afraid of an evolutionary God is more subtle than the other two. This reason has to do with social stability and social coherence, or to put it in a modern idiom – with the idea of Law and Order. In an indirect but powerful way, the conception of an absolute, unchanging and unchangeable God has been a hidden support to all those who have wanted to discipline the populace and keep it within the frame of law and order. The connection between an absolute God and the law and order issue is a subtle one. Let us first see how it worked in Plato.

Plato disliked Heraclitus and abhorred the idea of "everything flows". One of the important reasons for this dislike was Plato's contention that Heraclitus's philosophy was antithetical to social stability. Plato was an aristocrat. He believed that only certain chosen people could behave themselves. The populace had to be *controlled*. How can you build a social programme and social stability on a philosophy according to which everything changes? You cannot step into the same river twice. Or as Heraclitus put it himself: "We step and do not step into the same river, we are and are not." "No, this will not do," we hear Plato murmuring to himself.

Thus Plato set himself a task of creating *the ontological foundations* for social order and social stability. The invention of the Form (or Forms) as immutable, changeless and absolute, as underlying all visible and changeable phenomena, was a perfect solution to secure social stability, by relating it to the unchangeable and invisible ontological reality.

Plato was not the first to have invented a permanent, absolute ontological reality to serve as the foundation for and a controlling agent of social stability. Such a foundation is *assumed* in every religion which claims that God is the absolute ground of being.

Even if we were prepared to endow God with the attributes of changelessness and immutability nowadays – for the sake of social cohesion and social stability – it would be of no avail. Twentieth century society has been in the process of convulsive change. This change has been a part of a larger picture – of the whole universe around us changing.

The path to social cohesion and social stability in our times will not lead via the image of an absolute and wrathful God, who will bring order through His Iron Fist, but through making a new sense of the universe, including making a new sense of our own lives. And this will only happen when we become more connected within, more spiritually attuned, more responsible for the fate of the universe, more co-creative with the universe.

In brief, we cannot hope to achieve social stability by clinging to the image of an unchangeable absolute God who, as it were, wants his universe to be fixed and permanent, including human society. A new *Civitate Dei* must be invented as we are re-inventing the universe and finding a new place for God in it.

Let us attend to **the fourth reason** for our fear of an Evolutionary God. This reason has to do with the story of creation. In Christianity as well as in Hinduism, all creation springs from the ultimate ground of being. God and Brahman respectively are unchangeable, rock-like foundations out of which all fluid forms of life arise. It is emphasized that Brahman and God are grounds of *being*. They are the ultimate matrices of being. All other beings, and all forms of being, derive from the Ultimate Being. The whole emphasis is on *being*, on structure, on immutable elements.

Now, if we change our perspective and start to look at Creation from the vantage point of *becoming*, then traditional precepts and interpretations do not bind us any more. Thus, if we consider

Brahman and God as the *ultimate sources of becoming*, then we don't need to insist that they are permanent, fixed, absolute.

Underlying this entire discussion, there are two intellectual traditions clashing with each other. There is the Heraclitan tradition of becoming, on the one hand. This tradition seeks and envisages divinity in creative unfolding. Perpetual becoming is divine. How could it be otherwise? Evolution is divine. How could it be otherwise? Creativity is divine. How could it be otherwise?

And there is, on the other hand, the Platonic tradition, which insists on the primordial, imperishable bricks (Forms) out of which everything must arise. This tradition (which Brahman and the Christian God exemplify) is obsessed with structure, being, ultimate bricks as foundations of knowledge, and finally absolute subatomic particles which actually may be the last clutching of the straw by science in its vain pursuit of the immutable Jehovah.

On epistemological grounds, the point is this. We do not need ultimate atoms and immobile beings in order to understand the immensity of the universe and the depth of human soul. Karl Popper has shown that the entire approach of trying to understand science and all knowledge through irreducible bricks, structures, forms of being, *has failed*. The only way to understanding knowledge and reality is through understanding evolutionary becoming. This is the message which has been carried further in Thomas Kuhn's book, *The Structure of Scientific Revolutions. To understand is to understand how things unfold.* This is also the message of David Bohm's idea of Implicate Order. Thus becoming is the vehicle for the understanding of the reality of the physical cosmos and the reality of God.

When the issue is approached from the standpoint of God's omnipotence, and especially his creative power in action, the following argument can be stated in favour of the Heraclitan interpretation of God. Let us look at the very meaning of the phrase "The source out of which all forms spring and arise". This very expression suggests this source to be creative, dynamic, fecund, bursting with exploding energies. It simply follows that the ultimate source out of which everything springs must itself be open, creative, dynamic. *God is the source of all becoming, is part of this becoming!* The essence of the universe is creative becoming. The concepts of a perfect and a continually evolving God are compatible with each other. And who can *prove* with any certainty that an evolving God is less creative and perfect than the static and immutable one? *How can God be creative if he is inert and entirely static?* Rabindranath Tagore has put it thus: "God finds himself by creating."

The fifth reason for our fear of an evolutionary God has to do with the traditional meaning of the term 'perfection', and especially the sense of our psychological well-being as related to this perfection. The immutability of the perfect being is not only required for social stability, but also for our personal security. Within the Christian tradition we consider our personal lives well grounded if they are anchored to a permanent, immutable God. We think that we are much more secured in our personal lives if God is a changeless entity. But is this a right basis for our psychological security? Is it not the case that our personal security has always been the result of our own integrity, of our capacity to work on our inner selves to accomplish the peace within? Perhaps it is comforting to think that there is someone out there taking care of you. But perhaps it is also a bit foolish ... and irresponsible, as I will argue in Chapters 6 and 7.

The message of this chapter and of this entire book is: "Release yourself from your anchors and chains. Fly!"

Yet some would want to respond: "What do you mean by *fly*? We can only crawl."

Indeed, much too often we have crawled and got used to crawling. But this is not our destiny. We have crawled enough. We are not caterpillars. Evolution wants us to stretch our wings and fly – to our ultimate destiny, which is to become liberated, realized, enlightened, God-like. Only then will God be able to rest because the universe will have realized itself. Only then shall we fulfil our cosmic destiny.

Summary

God has created us so that we articulate his attributes and his nature properly. The more intelligent and mature the mind the more intelligent and enlightened is the conception of God. To have a mature conception of God, we need to mature sufficiently. We have no other means of knowing about God but through our minds. Even in our mystical states we are not mindless. On the contrary, our mind is then exquisitely attuned to the ineffable and thus capable of reaching the ultimate boundaries of our existence. *As our mind is, so is our conception of God.*

To savage people God has savage characteristics. To ecologically sensitive people, God possesses ecological characteristics. We have needed to mature to become aware that we are responsible for the fate of the earth, and for the healing of the earth. As our ecological sensitivity has grown so we have seen God greening and re-directing us to heal the planet and to take care of its creatures.

The image of the God of the Old Testament was born of concerns and problems different from ours. As our consciousness has changed – guided by new concerns and visions, of the interconnected and ecologically viable cosmos – so our image of God has changed, and thus the reality of God. The vacant place of the supreme authority over the supersensory world, of which Nietzsche and Heidegger speak, cannot remain empty forever. We have discovered that this place is not a reality external to us – though we often think about it in this way – but a reality of our inner selves. As we cannot live without our inner selves, so we cannot live without the reality of God.

Our God and our mind work in unison. As our mind matures so our God matures. As our consciousness matures and becomes ecological consciousness so our conception of God acquires the ecological aura, or, as we said before, the ecological dimension. The god of wrath becomes the god of ecological mercy, ecological healing, ecological redemption. "Our life is the creation of our mind" (Dhammapada). Our conception of God reflects the depth and sensitivity of our mind.

Perfection does not mean immutability. An evolutionary God is actually more perfect than the static one. Perhaps the principal fear of an evolutionary God is the fear of our terrible responsibility for realizing the divine potential within ourselves.

As we co-create with the universe, we find that the more creative our participation is the more creative the universe and God become. "If you do not expect the unexpected you will not discover it; for it cannot be tracked down and offers no passage" (Heraclitus).

To live in the universe of creative becoming – as contrasted with the epochs in which people were chained to fixed dogmas – oh, what a joy! To live with a creative and evolving God, oh, what an honour! To live in the universe of which we are the thinking mind and the seeing eye, oh, what a responsibility!

CHAPTER FIVE

Spiritualities in History

Technology as a Pseudo-Religion

The history of human kind is the history of wars. But equally, it is the history of new forms of spirituality arising. The Human kind is still base. Hence the wars. The human spirit is slowly refining itself. Hence the persistent search for new images of perfection, for new images of God, new forms of spirituality. Regardless of how many brutal wars we have seen in the past – and in present times (and wars are *always* brutal) – we still have a right, and indeed an obligation, to believe in the perfectibility of the human condition. Past atrocities do not nullify our divine potential, but only tell us that the road in front of us will be thorny – as we need to overcome numerous biological and psychological obstacles.

Besides, no spiritual tradition worthy of its salt promises an easy spiritual salvation, or an easy attainment of spiritual virtue. Each insists on painstaking work on oneself and each emphasizes that the spiritual attainment is a gift of the gods which must be deserved.

Only the technological optimists have promised an easy solution to our problems and an easy bliss. And look what a mess they have made of this world! Our present tragic dilemmas, including the Earth staggering under the blows of technology, are not the result of people being misguided by superstitious religious beliefs. On the contrary, these acute dilemmas are the result of a triumphant rational technology, which is actually being misguided by a wrong eschatology, the eschatology of consumption which promises bliss through the acquisition of objects. We have always known that this eschatology is a farce. Yet we have not protested sufficiently against its cancerous spread for somehow the glamour of technology has intimidated us.

Let us stop fooling ourselves that one day technology will deliver. It won't because it can't. Technology is not the kind of instrument that can bring enlightenment, thus fulfilment, thus genuine peace and happiness. Happiness without enlightenment is a trivial thing. We have seen it eloquently described in Huxley's *Brave New World*.

The more advanced technology becomes, the more sophisticated weapons of mass destruction it provides. The war in the Persian Gulf proved once more that the advancement of technology is tantamount to the advancement of new weapons and not to the promotion of peace and happiness. Our imbecility in our thinking about technology is remarkable. Yet there is a rational explanation for this imbecility. Technology has mesmerized us, has made us dupes to its glittering superficial aspects, has cast a *religious spell* on our confused minds. An allegiance to technology is the pseudo-religious responsibility of our times. For this reason technology has muffled our critical capacity to assess its devastating consequences.

Although technological culture is proving such a menace to our spiritual destiny, we should not think of it as an empty straw-man. It is clever, ingenious, resourceful and, above all, seductive. It has even created its own spirituality – of sorts. If spirituality is a recapitulation and a condensation of the human condition of a given time, then during the period of human stupefaction by consumption, *this ritual of consumption may become a form of spirituality*.

In brief, the glorious consumption of glittering gadgets is a form of spirituality for a technological culture. For in the glitter of the mechanically superior and electronically dazzling tools, gadgets and toys, the impoverished human condition of the technological age finds its apogee.

What about the saints of technological culture? These are the astronauts. There should be no mistake about that. They are treated with utmost respect, awe and reverence. They represent the highest achievements of techno-culture. Their individual prowess, intelligence, ability and performance should not be disregarded or diminished. For they are the best. Well, they represent the best of the technological mentality. They are the acme of technological theodicy. Their spirituality may be lacking – if measured by the traditional criteria. But that is another matter. And not the fault of "saintly" astronauts either. They embody and express the very best of technological culture. It is this culture that has crystallized the human condition in a specific way. It has created new criteria of excellence (technological prowess and efficiency), and it has created a new kind of saint – the most accomplished men who can manipulate the most accomplished machinery that has ever been produced in the history of human kind.

When you cannot become an astronaut, you can at least use teflon pans developed for the use of astronauts. You can use other glittering gadgets. You can consume glittering goods as part of the orgy-porgy of the Brave New World. The eschatology of consumption and technological spirituality support each other. Such is the state of the human condition in our present technological world.

Waging ecological peace means getting out of the spell of technological glamour; it means breaking the religious spell of technology. Waging ecological peace means resurrecting traditional spiritualities – in so far as they still possess a nourishing and living substance. Waging ecological peace means creating a new spirituality in the image of Gaia – alive, radiant and blossoming. This new spirituality is called ecological spirituality for good reason.

Spirituality as Independent of Images of God

Traditional spiritualities have survived through tales, through cultures, through religious systems of beliefs of various people. Many of these forms of spirituality are still nourishing, inspiring and guiding the people born within their realm. However, many of these traditional forms withered away, lost their capacity to nourish, became empty rituals – full of high sounding words, signifying little.

What is astonishing is the sheer variety of spiritualities and religious forms of beliefs. Now there are some schools that hold that ultimately there is *one* spirituality, *one* God, *one* religion. This is a noble sentiment, appealing to our sense of the unity of life.

However, let us be quite clear that we can uphold the unity of life while celebrating a variety of spiritualities, a variety of concepts of God, a variety of religions. Life flourishes in different forms; and so does spiritual life.

Various religions do not speak of the same God. Those who insist that they do may be indulging in a form of wishful thinking. Wishful thinking is an enormously powerful force within all religions. Sometimes this wishful thinking is called "begging the assumption".

It is then assumed that *God must be one*. Starting with this assumption, we simply impose the idea of *one God* on the variety of religious beliefs. Now, if one is adamant enough to see one God in all religions, there is nothing to stop one from doing that. Human reason is frail and powerful at the same time. It cannot prove anything with complete certainty. Yet it cannot disprove anything with certainty – especially propositions concerning the nature of God.

How can we ever settle whether there is one God underlying all religions or not? Do we need to settle that? This question is very important. Yes, but only to those who *assume* that there is one God. However, there are some major religions which do not speak of God at all as an underlying reality of their religious systems. Among them are Buddhism, Jainism, Zoroastrianism, Taoism, Confucianism.

The sense of the divine does not require any image of God. The sense of the divine and human spirituality had flourished long before religions, as articulated bodies of beliefs, came on the stage of history. What is important are not images of God – over which religious wars have been waged – but the conduct of our life, the beauty of our condition, our capacity to make the divine dwell in our lives, our capacity to release the divine from within. Inviting the divine to dwell in us, releasing the divine from within, creating the divine through our lives are aspects of the same process – of evolution making itself more divine through us.

All living beings came from the same plasma. Yet as evolution has unfolded, this plasma has articulated itself in different life-forms. As a result humans are not horses, bumble-bees are not swallows, coconut trees are not oak trees. Let us rejoice in this variety. Let us celebrate the difference. Variety is in the nature of life. Homogeneity is the first step to death.

All religions came from the same impulse: to reach beyond, to actualize this mysterious seed in us which makes us genuinely human, that is divine. In a word, the imperative of all religions is to transcend. But various religions have chosen different paths for this process of transcendence. Hence the variety of heavens, the variety of nirvanas, the variety of spiritual paths. Let us celebrate this variety, for God loves variety and expresses his creative energy through variety.

Let us ask ourselves why some religions so persistently cling to the image of one God (or to the idea of the Absolute Ground of Being)? There is a historical and psychological reason for that. The roots of many major religions are ancient. In dreaming of a perfect reality, within which all hardships of the immediate imperfect reality are redeemed, those ancient people imagined a paradise which was supposed to exist in still more ancient times, and which became the Paradise Lost. This was transcendence running backwards. Those ancient people did not have much idea of the awesome beauty of creative evolution and how it transforms all forms of life. Their mind was simply not ready to entertain the idea of evolution as embracing all, and also the idea of the divinity as the process of mind-making and ultimately God-making. Instead their consciousness, and thus their spirituality, crystallized in another way. They condensed all divinity and grafted it onto an abstract point, existing outside themselves and outside the life process. They called this point 'Brahman' and 'God' respectively. They sanctified this point, made it absolute and unchangeable.

Then they became attached to their conception, particularly as they started to weave around it complex patterns of thought, and complicated rituals, as well as sophisticated cultural designs – which become forms of life. A tradition was built, then perpetuated for centuries and millennia.

Once you are born in a tradition, you do not question it. For you are it. No wonder it is comfortable for the Hindu, and so very 'Natural', to believe in Brahman as the absolute ground of being. And for the Christian to believe in a personal God, who resides outside the human universe. But take a Hindu child from the cradle and give it to a Christian family and this Hindu child will be comfortable with the Christian God. Take a Christian child from the cradle and give it to a Hindu family and it will grow up a Hindu and will find Brahman and all precepts of the Hindu tradition perfectly natural. This just shows that those fundamental beliefs, which feel so 'Natural' within a given religion, are *culturally conditioned*.

No doubt, the crystallization of human spirituality around those novel perspectives – Brahman and God respectively – was a brilliant departure, but the first approximation, not the last word on the subject of divinity and God. It would be absurd to think that those ancient people who shaped major religions got all their answers right; or to assume that God does not want to speak through us, or would not wish to enlist our participation in making the world a more divine place.

It is difficult to question and revise the ideas which your culture has laid as immutable and written in heaven. One of these ideas is that of absolute laws – as related to the absolute ground of being; or as expressing themselves as aspects of the absolute reality. But again, the idea of absolute laws was the first approximation. In fact, there are no such laws. (If there are, as I said before, we are not intelligent enough to know them.) We have to learn to live with change, and be humble about our claims concerning absolute knowledge.

The Buddha was absolutely clear on this point. The only immutable law, he said, is the law of change. In one of his discourses the Buddha spoke thus:

> There are no fixed laws and rules in terms of which everything is to be evaluated. This, monks, is the Middle Path … the adoption of fixed rules of discipline will enable man to attain freedom only if everything in the world, including human life, is governed by fixed laws. But I see no such fixed and unalterable laws in the world. Not seeing such invariable laws and seeing things as they have come to be conditioned by various factors, *I declare the world to be in a Process of Becoming.* How can I prescribe inviolable laws of discipline for the monks?

The poet Shelley once said that "Poets are the gigantic mirrors in which the future casts its shadow upon the present." Similarly with our longing for perfection and our attempts to pin down God within the confines of our present reality. Following Shelley's idea, we can say that *God is the gigantic light of the future, and spiritual people are those who develop inner mirrors to catch glimpses of this light within their own souls.* The longing for this light is inevitable. The inadequacy in receiving it is obvious. Yet being forms of life which are already partly illumined as the result of the process of transcendence, we are compelled to reach out. We cannot help but seek God. For in this seeking we are following our destiny, we are following our inner potential. Thus to seek the absolute does not mean to accept the idea of absolute laws, or the absolute ground of being, which is called God. Seeking the absolute may be an infinite process. Let us not trivialize it by assuming that we know what the absolute is. To reiterate, the sense of the divine does not require any image of God, any postulation of the absolute ground of being. The divinity of life can be cherished, upheld and perfected without the notion of absolute laws, without the notion of God.

Spirituality and rationality do not exclude each other. Rationality is divine too: to *conceive* of God and of divinity, to *understand* the difference between the divine and the base is part of divine understanding. All understanding is a gift of heaven. Thus ecological religion makes us aware that the whole transcending process of reaching divinity is natural and rational at the same time.

This whole world is in the glory of God, is pervaded by God, is realizing God. What could be simpler and more rational than that? What could be more mysterious and more inscrutable than that? And aren't the two – the rational and the mysterious – two sides of the same coin? It is quite rational to accept mystery as part of our endowment and part of the world we live in. It is quite mysterious that rationality works and that we can understand anything. Let us not try to figure out *now* what God is, and what ultimately our divinity will become. We have the next five millennia for this purpose, and perhaps the next fifty million years.

There is a sense of hubris and more than a touch of arrogance among people who seem to know so well what the Brahman is, or what God is. (Although often they say they know nothing of the Brahman or of God, they go on endlessly explaining the nature of each.) The attainment of deeper spiritual knowledge should signify a knowledge of the limits of knowledge. And aren't the limits of our knowledge so obvious when we seek the knowledge of God and of the Divine?

Different Spiritualities as Different Ways of Articulating The Divine

When we look at the variety of religious beliefs and spiritual practices, we are astounded at how varied are the manifestations of the human spirit, and how varied have been the flowers of blossoming divinity. Yet from the evolutionary point of view, this should not surprise us at all. The richness of life is the flower of an ever-transcending evolution.

Now, these various forms of life, and of spirituality, do not spring at random. They blossom in distinctive niches. To understand the history of the spiritual life of humankind is to follow and understand these niches. Each form of spirituality contains several layers, or is made of several components. Among these components the most important are:

1. The historical context of a given spirituality/religion
2. Its story of creation
3. Its postulated ground of being
4. Its source of divinity
5. The concept of redemption or of heaven
6. The ideal of responsibility
7. Man's relationship to nature
8. The overall concept of man
9. The vehicle of transcendence

If we mention some of the major forms of spirituality, which have significantly shaped the human psyche, such as:

Buddhist Spirituality
Christian Spirituality
Confucian Spirituality
Hindu Spirituality
Islamic Spirituality
Judaic Spirituality
Taoist Spirituality

then it is clear that a true picture of human spirituality in history could only emerge after we have studied each of the major spiritualities – vis-à-vis the others. If we reconstructed human spirituality painstakingly, we would end up with a magnificent tree whose branches go in so many directions, yet all trying to touch the heavens.

Our purpose is not to write a history of human spirituality. Therefore we shall not attempt to reconstruct each spirituality in its historical setting. Instead, we shall sketch some parallels showing how different spiritualities were bound to create different kinds of psyche; and how different people were bound to articulate spirituality along different lines.

Different religions shape the phenomenon of man in a different way, create, in fact, different kinds of humans. These different kinds of humans interact with the world differently. In interacting with the world differently, they create the world differently. We live in a participatory universe. We are co-creating with the forces of the universe – divine or otherwise.

If we take Hinduism and Christianity for instance, and look at their respective origins – what an immense difference. Hinduism grew in the valleys of plenty. It grew slowly without any rush or external pressures. The Vedas, those immortal hymns that laid the foundations of Hinduism, were poetic songs. They are full of profound wisdom but they are also rhapsodies expressing the sweetness of life.

Thus the Rig Veda enchantingly sings:

> For one that lives according to Eternal Law,
> The winds are full of sweetness.
> The rivers pour sweets.
> So many plants be full of sweetness for us.
> Sweet be the night and sweet the dawns.
> Sweet the dust of the earth.
> Sweet be our Father Heaven to us.
> May the forest trees be full of sweets for us.
> And full of sweetness the Sun,
> May the kine be full of sweetness.

In contrast, Christianity was born out of harshness. The ten commandments are harsh prohibitions. The Bible itself is full of harsh stories. Some of them are horrendous. The children of Israel have had it rough, the Exodus and all. The crucifixion of one of their best sons is the continuation of the harsh story.

Between the time that the *Vedas* were composed and the time that the *Upanishads* were written – those sublime metaphysical treatises in which the poetic insights of the Vedas were given a coherent and discursive form – some five or six centuries passed. This was a period of a slow and exquisite refinement of the foundations of the Hindu mythology, of the world view based on the idea of the harmony between the Brahman (the ultimate ground of all being) and the Atman, the individual Self which longs to be united with the cosmic self (Brahman). The Hindu psyche and the Hindu way of life were slowly chiselled out in the caves of the Himalayas and in the luxuriant forests with which India was covered at the time. All in slow pace, and without any rush. This slow pace is still part of the Indian psyche.

In contrast, Christianity seems so often to have been in a rush, in a state of emergency. When Constantine the Great *declared* Christianity to be the official religion of Rome, in 313, this was the first emergency. When, after the collapse of the Roman Empire, poor St Augustine had to deal with so many new legal matters, for which he was unprepared, this was another emergency. And so the pattern was set – from emergency to emergency. Each synod was in response to an emergency. Ideas and beliefs did not have time to mature gradually so that the religion could grow organically, as part of the breathing space of the religious community. Instead edicts were proclaimed and dogmas were announced. Christianity is a religion ruled by dogma.

This rush, emergency, restlessness and the combative nature of Christianity can be seen in the crusades, can be seen in the continuous battle between the popes and the emperors; and above all in the continuous acts of excommunication of those who diverged from the dogma. And of course in the institution of the Inquisition.

Somehow the Hindus were settling these important matters differently. When Eadi Shankara decided to bring the land of India back to the fold of traditional Hinduism, he went through the entire land and debated with everybody. No wars, only persuasion, accomplished by the strength of one's being and one's ideas.

One wonders whether our restlessness, that is to say the restlessness of western people – brought within the precepts of Christianity, our combativeness, our continuously rushing nature – might have to do with the characteristics of the religion which has shaped us. After all, Christianity did not have a sweet quiet time (like Hinduism did) for its maturation. It has always been on a crash course, always fighting, if not against external enemies then within its own ranks.

We westerners are so often struck by the passivity of the Indian soul which most of the time we take to be a sign of inferiority – in comparison with our robust activeness – without realizing that their passivity may be a residue of the tranquillity of their religious roots, while our activism may be a residue of the restlessness of our religion. All of this should not be construed as a criticism of Christianity or a praise of Hinduism. Rather it should be viewed as a reminder that different religions and their respective spiritualities shape the phenomenon of man so differently.

Another characteristic feature of Hindu religion is its timelessness. Somehow time has a different dimension in the Hindu tradition
– it is not so pressing, not so important. Hence chronology is not
that important within the Hindu culture. We don't know with any
certainty when major events took place, when the Vedas and the
Upanishads were written. At times estimates vary by a margin of
millennia.

This sense of timelessness is linked with the sense of eternity
which pervades the Hindu mind. Add to this the presence of the
Himalayas, which are always on the horizon. The Himalayas have
always been a permanent temple in which the Hindu spirit has found
a shelter. When your gaze is permanently fixed on the far away
horizons of the Himalayas, and when you live in the space of
timelessness, if not of eternity, then you are bound to create a cast
of mind within which the mystical longing for the transcendent
becomes a part of the natural breathing of the soul.

This is not to say that the Christian tradition has not had its own
form of mysticism and an elevated sense of transcendence. But the
mystic tradition – within Christianity – and the understanding
through the heart, have been gradually suppressed in favour of the
legalistic tradition and the understanding through the mind – if not
through the dogmas. (More will be said on the mystic tradition of
Christianity, and particularly on the anticipation of ecological
spirituality within Christianity, in the chapter on St Francis.) Let us
make another comparison, this time between Christianity, Hinduism and Buddhism. Let us compare their dominant symbols. And
let us see how these symbols have shaped the psyche of the various
people living within the energy fields of these symbols.

The central symbol of Christianity, particularly as seen in all the
Churches, is Jesus carrying the cross to Golgotha. The symbol of
Christianity is the cross. Through this symbol we are guided to view
life on earth as misery, as a valley of tears. Now, it is true that Jesus
preached the ideal of Life Abundant. It is true that there is another
symbol within Christianity – that of Resurrection. The symbol of
resurrection, as a continuous becoming of life, as a continuous
renewal, as a continuous miracle of life re-born, would have been a
very powerful and very appropriate one for our times. Yet it is not
this symbol that dominates Christian consciousness. It is the cross.

What a contrast to the Hindu conception of life! The most
striking and powerful symbol of Hinduism, the dominant symbol,

is that of Dancing Shiva. It is a symbol of continuous metamorphosis, of the ceaseless becoming of life, of essential fluidity, and also of the fecundity of life. There is something fascinating and mysterious in the conception of the universe conceived in the image of Dancing Shiva.

The dominant symbolism of Buddhism is the Buddha sitting serenely on the lotus flower. The lotus flower itself came to signify the quality of the Buddha – the inner peace of mind, which is a precondition of real well being, and of happiness. The symbolism is simple, but powerful and universal.

Now these dominant symbols have profoundly shaped the psyche of the respective people and their corresponding spiritualities. It is almost inconceivable to think what the Christian soul would have become had it not been shaped by the symbol of the cross. It is equally inconceivable to think what the Buddhist soul would have become had it not been shaped by the image of the Buddha sitting on the lotus flower. It is different with Hinduism which is so colourful and relies on so many images.

Symbols of religion are artifacts of man, not a divine gospel sent by God. The same divine truth can be revealed in many different symbols. On the other hand, the various symbols can be seen as articulating different aspects of the divine. We cannot judge the value of a religion by the value of its symbols (what is the value of a symbol anyway?). Each set of symbols articulates a different aspect of the human condition, was born in different circumstances, induced by different historical necessities.

The discussion in this section was meant to make us aware how different the ethos of various spiritualities can be, depending on the historic circumstances in which these spiritualities were shaped, developed and moulded. Yet the form of a given spirituality is not a haphazard matter. We have been arguing that spirituality – any spirituality developed in history – is not absolute in character, is not fixed in heaven and therefore timeless. It needs to be emphasized, on the other hand, that spiritualities are not ephemeral and subjective phenomena.

There is a fit between the total circumstances of a given people and the kind of spirituality they develop. The direction in which the human psyche is developed, within a particular culture, or a specific religious tradition, is precisely the form of articulation of the human condition. There is nothing subjective or relativistic about the form of spirituality that prevails within a given culture. Spirituality and culture fit each other like a glove that fits the hand. This is so within cultures which are vibrant and coherent, and within religious traditions which are alive and nourishing.

When this fit occurs, spirituality is not just a mirror reflecting the main features of a given culture but is a roof crowning the structure of which the supporting pillars are: cosmology, consciousness, values, eschatology. Let us explain the matter in some detail. Within coherent worldviews and in vibrant cultures, there is a beautiful harmony between the cosmology of a people and their consciousness. They reflect and co-define each other. Values and ultimate goals (eschatology) are part of the picture. Values so often articulate consciousness and are themselves guided by eschatology. Spirituality embodies and summarizes it all. By reading in depth the meaning of spirituality, you can decipher the underlying cosmology and the prevailing eschatology. By reading in depth the underlying consciousness of a people and their values, you can deduce their spirituality. The fit and connections are there. But it requires a subtle mind to see it all.

Now, the trouble starts when you try to change one element of this fabric without being fully aware how all these elements fit each other and support each other. Let us take some examples. Many westerners are dissatisfied with the meaninglessness of their lives and the spiritual vacuum brought about by the materialist lifestyle. As a result many have embraced eastern gurus and sought salvation in eastern forms of spirituality, which seems fine. But it isn't. For the point is that while they deliberately cultivate a Hindu or a Zen-Buddhist spirituality, *they are stuck in the old mechanistic consciousness. In their day to day living they adhere to technological values and to the eschatology of consumption.* Thus there is a clash within the individual. Sometimes eastern gurus subtly or deliberately bring in the eastern cosmology, eschatology and mythology as a part of the supporting picture. But most of the time it does not work either. Images and assumptions of eastern cosmologies (with the exception of Taoism and Buddhism perhaps) are remote from western consciousness.

Let us consider another example. Some other people, equally alarmed with the spiritual devastation of rampant materialism, have chosen to seek a spiritual renewal by going back to the spiritual roots of their own tradition – Christianity. But they have often found the outcome unsatisfactory. The roots of Christian spirituality lie in the Biblical story. This story is no longer satisfactory to us. We need a new story, as Thomas Berry has emphasized over and again, a new story of creation, a new cosmology within which our spirituality can find a focus and out of which it naturally grows. Cosmology is a home for spirituality, a natural habitat out of which it grows. The old story is simply not credible. We just simply cannot believe it.

Some others still, devastated by materialism and unable or unwilling to search on their own, give themselves to fundamentalism as a path of spiritual renewal. But such a path signifies relinquishing our responsibilities. You cannot find spiritual liberation by giving away your soul to be manipulated by others.

Let us take yet another example. So many people are alarmed by our environmental plight. The awareness of this plight has brought about the movement called environmentalism. Within this movement, environmental ethics has been created. Yet frequently this ethics is disconnected from any spiritual quest, and is based on some kind of calculus, called cost-benefit analysis. In the short run, this kind of ethics may help us a little, by making us at least aware of the devastations taking place – and how we can devastate less. In the long run such an ethics is a futile enterprise as our new values do not have any firm foundations. Cost-benefit analysis is not a foundation for any ethics. For environmental ethics to succeed, we need a new cosmology, departing drastically from mechanistic cosmology, which after all is the basis of our rapacious materialism; and we need a guiding framework of a new spirituality. In short, environmental ethics (which I prefer to call ecological ethics) in order to be this ethical instrument which sufficiently inspires and properly guides us in healing the earth and ourselves, must be based on eco-cosmology and on ecological spirituality as its foundations.

Summary

Spirituality is intertwined with our world view, with our values, with our goals, with our sense of destiny, with our concept of heaven, with our idea of salvation.

The form of spirituality of a culture grows organically out of this enormous tree of life which a given culture cultivates. Spirituality may be considered the sap which nourishes the whole tree. When this sap is weak and devoid of vital energies, the whole tree withers. Each tree of life, cultivated within different cultures, requires different spiritual energy as we have seen from a brief examination of different spiritualities.

Our western worldview is in pieces ("Tis all in pieces, all coherence gone"). Our values are in a shambles with primitive relativism and rampant nihilism running amok. The meaning and purpose of our life are constantly put to question. Past spiritualities are enfeebled and unable to nourish us spiritually. Thus we seek a new spiritual nourishment, and a new purpose in life. Often we run frantically and incoherently, once in this direction – seeking a spiritual renewal; once in that direction – seeking new values. But this is to no avail. We must renew the whole tree in order to be healthy and whole again. Thus we must reconstruct our spirituality while we are reconstructing our consciousness, our cosmologies, our values.

Only when we have redefined our cosmology and our consciousness will we feel comfortable with our new ecological spirituality. Ecological religion is thus a whole new magnificent tree of life. Let us emphasize, ecological religion is not painting green the edges of past religions, but making sense of the new story or creation, practising reverence for life in our daily affairs, finding a new transcendent purpose in our individual lives, establishing a new idiom of cooperation with God, whereby we see ourselves as co-creative participants in the unfolding story of creation.

CHAPTER SIX

The New Physics Embraces the Divine

What to do with Science?

One of the problems which faces every religious person and every religious institution is: what to do with Science? During the last three centuries science has grown at an astonishing pace as an autonomous realm, proclaiming its truths in all domains of human life, while being accountable only to itself. So important has science become as a cognitive force that all other fields of human intellectual activity, including religious thinking, have felt obliged to be accountable in the court of science. And hence the problem.

Classical science, derived from Newton's mechanistic paradigm, has increasingly separated itself from its religious roots and declared, if only implicitly, that religion is spurious and that science can explain it all. The chasm has grown very deep, actually to the point that a sharp dichotomy has appeared: either you are rational and you believe in science, or you believe in religion and God and therefore you are irrational. This chasm, perpetuated by secularism as the leading ideology of the post-Renaissance era, led to the emergence of a pugnacious materialism, which in due time produced its one specific embodiment – Marxism, which was a characteristic exemplar of a new kind of secular religion based on faith in Reason, Science, Technology, Progress, and the Redemptive Role of the Proletariat.

In brief, until the second half of the twentieth century it was difficult to be a rational person, to believe in science and at the same time to be a religious person. Because of the enormous authority of science, the Churches themselves adopted a low profile as they were unable to stand up to the authority of science.

The separation of science from religion was viewed by many as unfortunate. Thus numerous attempts were undertaken, from the second half of the nineteenth century onwards (and even earlier if we remember Auguste Comte's monumental work on the system of positivist philosophy), to aim at reconciling science with mysticism, or science with religion. The Indian thinkers were as prominent in these endeavours and as ingenious as were Western ones. But ultimately, all these attempts were futile. For the simple reason: you cannot reconcile mechanistic science – which believes in the existence of matter only – with transcendental beliefs in non-material entities such as God, divinity, meaning, grace, freedom.

Only in the second half of the twentieth century when science, especially the New Physics, has dramatically begun to change its nature and to recognize new kinds of entities, therefore enlarging its universe very significantly, has the reconciliation of science with religion become a serious prospect. However, at this juncture of history when science has broadened its universe so considerably, we do not need to talk about reconciliation any more, for science itself has become a religious realm – as I will attempt to demonstrate in the course of this chapter. There is no longer any problem with science, as the New Physics embraces the divine and is, in fact, at the forefront of a new religious thinking.

The New Physics and its Universe

The New Physics has been weaving for us a new story of creation. It has postulated events and phenomena staggering in their implications. The New Physics can be called a new metaphysics; but also a new theology. Its theological implications are staring at us – if we look deep enough into the kind of claims the New Physics is making. These implications, however, are not so easy to articulate, especially in one coherent body of a religious discourse. For the time being, we have glimpses and partial insights, not yet a comprehensive synthesis.

The scope of the term "the New Physics" is not precisely determined. By "the New Physics" I mean (along with many others) the post-Einsteinian physics, quantum theory and its implications, sub-atomic particle physics and its recent implications, the new cosmology and its implications; but also such theories as David Bohm's of Implicate Order and Ilya Prigogine's of Dissipative Structures. In short, the New Physics is a host of new theories and hypotheses which give us an altogether different perspective on the universe – its origin, its dynamics, its order, its evolution, its meaning, including the meaning of our own life in the new universe.

Let us be quite aware that what the new knowledge unveils to us is really a new universe. We are not merely confronted with a retouching of the old picture of the Newtonian universe. The picture of the universe is so new that *we do not know how to look at it*. Literally. Our ways of perceiving are conditioned by our knowledge and by our mind. If it is not in the mind, it is not in our knowledge, and consequently is not in our eyes; that is to say not in our capacity to make a coherent sense of things on the basis of visual clues. Our perception gives us only visual clues. These clues must be encoded and reconstructed within some coherent patterns. These are the patterns of accepted knowledge.

Because the patterns of accepted knowledge, until recently, have been so dominated by the mechanistic paradigm, we simply could not see certain things in the universe. We still can't – in so far as we allow the old patterns to determine for us what we consider to be real. The tremendous explosion of new knowledge during the last fifty years calls for the creation of new patterns of interpretation. These new patterns, when they become coherently woven together, will amount to a new world view, a new coherent and articulate picture of the universe. Only then will our new perceptions be properly guided and given a sufficient support; and only then will our new perceptions enlighten us in our quest to see the universe in a new way.

Among the powerful insights of the New Physics should be distinguished at least the following:

1. *The renunciations of the claim to objectivity*; sometimes partial, sometimes total. Newtonian physics assumed that the world is transparent and that it can be photographed (so to speak) in our scientific theories. These theories – which are a mirror we put to reality – give us a precise and adequate picture of what reality is and how it behaves. This is the essence of scientific objectivity, namely

the claim that we can objectively grasp nature (and the behaviour of the whole universe) in our scientific theories. This claim to objectivity has been undermined in many ways during the last forty years. First, Quantum Mechanics has decidedly proved that we cannot measure objectively the position and momentum of a sub-atomic particle, for while measuring it, we interfere with its position. This insight has been generalized. It is now seen and recognized that our entire knowledge (and the human mind built into it) interferes with the state of things as we attempt to "grasp" them through our theories. As we have devised more and more subtle and in fact incredible theories – within particle physics and the new cosmology – we have increasingly become aware that our "fantastic" theories and hypotheses are powerful shapers of reality (we should really say of the new reality we have been devising) and so much so that we are more and more convinced that reality is a hostage of our imaginative theorizing.

What we perceive and find depends on the power and ingeniousness of our hypotheses. Some of these new hypotheses have been so much beyond the common sense, so incredible, in fact, that they border on the realm of fantasy; nay, they are an expression of some kind of fantasy; except that when confronted with reality, or should we say, when imposed on reality; or should we say even more precisely, when used for an imaginative reconstruction of reality – these new hypotheses (seemingly from the land of fantasy) result in new and plausible, sometimes audaciously beautiful, patterns of interpretation. Thus reality obliges and can be moulded in various ways. *Reality is not a stable rock which we photograph in our theories, but malleable clay which allows us to form it in various ways – through our imaginative theorizing.*

Thus objectivity is out. And there is no quibbling about that. What is *in* is the principle of cognitive participation, the realization that *whenever* we attempt to describe reality, we actually interfere with it – by moulding it through powerful and usually invisible tools and filters which our theories and linguistic descriptions are. Thus the principle of *cognitive participation* is replacing the principle of objectivity.

2. *A new theory of mind.* Woven into the tapestry of the principle of cognitive participation, there lurks behind it a new theory of mind. Not only lurks behind but is clearly visible, if we look deep enough. What is new about this theory of mind? We could call this

new mind, the participatory mind.* The participatory mind is in sharp contrast to and actually replaces the old empiricist concept of mind, within which mind is conceived as tabula rasa, a white sheet of paper on which experience writes its designs. At best the empiricist mind is a passive mirror which reflects the static shapes of reality 'out there'. The two go together: reality conceived as a static rock and mind conceived as a passive photographer of this reality. And the two have been transcended. The mind is increasingly recognized as a shaper and moulder of reality, an important participant in the project called the description of reality. Let me quote some physicists on the subject.

> The notion of reality existing independently of man has no meaning whatsoever.
>
> Bernard D'Esspagnat

> The universe does not exist 'out there' independent of us. We are inescapably bringing about that which appears to be happening. We are not only observers. We are participators. In some strange sense this is a participatory universe.
>
> John Archibald Wheeler

> When understood through the implicate order, inanimate matter and living beings are seen to be, in certain key respects, basically similar to their modes of existence.
>
> It may indeed be said that life is enfolded in the totality and that, even when it is not manifest, it is somehow 'implicit' in what we generally call a situation in which there is no life.
>
> David Bohm

3. *A new perception of matter.* Within Newtonian physics, matter is regarded as this dumb inert stuff out there – moving, like billiard balls, by push and pull. When the co-creative powers of the mind are recognized, the entire approach to matter becomes more subtle and more sensitive. The simple truth is: the more sensitive the mind, the more sensitive becomes matter handled by it. The more intelligent the mind, the more intelligent matter becomes. And conversely, the more obtuse the mind the more obtuse is matter. Things reveal their nature in interaction with other things. There is

*For further discussion of the participatory mind see: H. Skolimowski, *The Theatre of the Mind.*

no such thing as crass matter. It is a property of matter that it can become a human person. "Things in themselves" (Immanuel Kant) do not make sense, for things are never in themselves, but always interacting with other things, always a part of the participatory process into which mind is inherently woven. What other secrets and attributes we shall find in matter will depend on our intelligence and the sensitivity of our mind, which will be able to tune itself into inner songs of matters.

Let me be now more succinct in enumerating other significant aspects of the New Physics.

4. *The recognition of hidden forms of order*, beyond and above the physico-chemical order of things. An example of these hidden orders is Bohm's Implicate Order.

5. *The recognition that the entire planet Earth is a living organism* – the Gaia hypothesis, of which much has been written in recent years.

6. *The recognition that the entire universe may be one living organism.*

7. *The recognition of the Anthropic Principle*, which postulates that the universe was so made as to bring life (and intelligent life!) about.

8. *The recognition of awe and mystery in the universe.* After so many exquisite hypotheses have been framed – testifying to the extra-ordinary richness of the human mind and of the universe itself – scientists are no longer inclined to believe that the universe is transparent; nor are they inclined to think that scientific knowledge is nearly completed and that we know nearly all that is to be known of the universe. Instead, they have learnt to appreciate the sense of wonder while exploring the universe. They are inclined to celebrate the awesome mystery of the universe – its depth, its profundity. This has led to a new modesty about what we know about the universe, and also, on another plane, to an almost *religious feeling* in our approach to the knowledge of the universe, which is no longer conceived to be "about the dumb stuff out there", but is viewed as a wondrous and splendid opus of which we are integral and indispensable parts.

At this point the approaches of the New Physics and of traditional religion truly converge. They become aspects of each other. In brief, science is again touching and exploring the glory and majesty of God. The sense of mystery and of awe, which the new mind of science exhibits, has been beautifully expressed by Einstein, who said: "The most beautiful thing we can experience is the mysterious. It is the source of all true art and science. He to whom this emotion is a stranger, who can no longer pause to wonder and stand wrapped in awe, is as good as dead: his eyes are closed."

Now, the various phenomena and insights of the New Physics which we have touched upon reveal not only a new cognitive situation, namely that we are making an altogether new sense of the universe in the intellectual sense, but also reveal that we are in *a new theological situation* – as we have to rethink the nature of God, the nature of the divine and our own place in this new wondrous universe.

What should not escape our notice is the fact that while puzzling out the nature of the universe and the place of our mind in making the universe, we have been, at the same time, puzzling out the nature of divinity and of God. *We are confronted with one stupendous puzzle: how to make sense of the combination of God/Universe/Knowledge/Mind.* Each part of this puzzle contributes to the meaning of the other parts.

From Rigid Deterministic God to Creative Ecological God

As the universe unfolds and the human mind unfolds, the panorama changes. New aspects are revealed; the old aspects, which were once thought of as unshakable and permanent, show themselves to have been of transient importance. As our journey continues, new things acquire importance. The depth of the universe and its dimensions change. Even God is being looked at in a new way. And therefore He/She begins to reveal new dimensions, unperceived before. We have already established that theology and physics are interconnected on the level of ultimate assumptions of a given cosmology.

I shall now argue that Newtonian physics was a form of codification of the basic tenets of the Judaeo-Christian Theology. Newton wanted to re-establish the glory of God. A perfect God could not have created an imperfect disharmonious universe. Showing that all phenomena in heaven and on earth obey the same immutable set of laws was a demonstration of the perfection of God – through the harmony of his creation. Let us be supremely aware that *it was the Christian God that inspired Newton's design.*

The biblical story of Genesis subconsciously guided Newton's conception of the physical universe. The world of Newton is rigid and deterministic; and absolutist in its nature. And so is the concept of God in the Old Testament. *Rigid deterministic God creates the world in his image.*

Newtonian Physics is in the image of Jehovah of the old Testament. Protestant Ethics is an extension of both. The three of them: the theology of the Old Testament, Newtonian Physics and Protestant Ethics spell out our essential unfreedom. We are crushed on three sides: by the will of the inscrutable God, by the deterministic laws of physics, and by the paralysing grip of morality according to which you slave and slave and slave here on earth, and your rewards ... may be in heaven.

This essential enslavement ends with the New Physics and the New Theology, each of which sheds a new light both on the nature of the physical reality and the nature of God. The New Physics also outlines the new boundaries of the phenomenon of man.

The God of the Old Testament – detached, full of wrath and beyond our reach, and the physics of Newton – objective and cold, share the same matrix, which is *the matrix of confinement.* They also make the human being small, lost, insignificant. We are pitiful against the inscrutable God-Father, Jehovah. We are equally pitiful against the inexorable laws of Newtonian mechanics.

In short, the old physics and the old theology spelled out confinement. At the root of this confinement was the separation of God from man, and the separation of the physical universe from the human universe: the outer and the inner were galaxies apart.

The assumption behind my arguments is that we have been too much influenced by the Old Testament while developing physics and shaping other forms of knowledge.

Newtonian physics was the product of the pre-evolution era. Hence its findings are expressed in absolute terms. The conceptual blinkers imposed on us by the bequeathed theology were just too great to allow Newton, or anyone else for that matter, to contemplate the world along the lines of the twentieth-century New Physics. Thus, *there is the theological undercurrent which runs parallel to the visible story of classical physics.*

Only after we have opened our minds to permit ourselves to see the world not as a rigid framed picture – framed according to the pre-established God's design – only then could we consider the story of the world to be a running film; and only then could we contemplate the species as evolving. Before that time the conceptual and theological blocks did not allow us to *imagine* differently, to *conceive* differently, to *perceive* differently. Imagination, conception and perception are all connected. Darwin began to perceive. The theological blocks in his time were already loosened. It is important to emphasize, however, that Darwin was not Mr Evolution. He rendered compellingly only one aspect of evolution – the evolution of the species through adaptation, biological evolution.

We are now beginning to see a larger picture emerging, the picture of the Western mind getting out of the predetermined frame and venturing to see the vistas of the world evolving. Let us recapitulate the story. We first assumed that the earth was created as we see it at present. We simultaneously assumed that all species were created as they appear to us now. We furthermore assumed that all physics is given to us as a package, once and for all. We simultaneously assumed that God represented a fixed frame, firm and immutable forever.

We have now discovered that physics (and by implication all science) is not given to us but continually created by us. With the understanding that *everything* is evolving, we are also slowly beginning to perceive that our God is evolving as well.

The New Physics and the New Theology spell out liberation. The name of this liberation is creativeness within the compass of evolution unfolding, and within the compass of our unfolding understanding.

The New Physics is not a local affair of phycisists or even one confined to the vicissitudes of science. Especially it is not the story of the old paradigm breaking down, and the new paradigm having a hard way of establishing itself; thus it is not a story limited to science. *It is something far more significant*. It is the story involving the third stage of evolution, of recognizing that science, and all our cognitive products, must be viewed as a part of the evolving film as well; it is the story of the recognition that we are both viewers and the makers of the film.

Furthermore the New Theology, which I call Eco-theology, the *creation*-oriented theology, as distinguished from old Christian theology which is *redemption*-oriented, is again not a matter to be confined to the theological hair-splitting of the Churches. Neither is it a matter of breaking down the old theological paradigm – that is after Nietzsche announced that God is dead – but is a matter of recognizing that *all* is evolving, including God. Originally God was a fixed reality, the frame that held everything firm. Now with evolution conceived as a running film, we no longer have any fixed frames. *The time has arrived to look at the evolving nature of the Framer*. This is what I call the fourth stage of evolution, one that encompasses the theological or the transcendental dimension as well.

Finally, the New Physics and the New Theology are not two separate developments, to be understood in their own specific terms, but parts of the same larger process of becoming. Thus we can say that human knowledge and human spirituality (including religion and God) are subject to the same all-pervading evolutionary flow. We have to have the courage of accepting the ultimate consequences of the evolutionary perspective.

The Newtonian paradigm, as the ultimate frame, was already profoundly in doubt at the end of the nineteenth century when such phenomena as radiation were clearly outside the frame. With Einstein we tried to mend the old frame and maintain that the bigger frame of Einstein is correct and that it contains in itself the Newtonian picture. But the new enlarged frame was problem ridden from the start: it suggested many counter-intuitive notions; besides as a frame it had fuzzy edges. We have never been *comfortable* with it. Our common sense has been shaped by the Newtonian language and its concepts. Still with Einstein we have a frame – a bit open ended and not quite precisely delineating its territory, but something to hold on to.

With the New Physics we are in a new situation. We don't have the frame given to us, which we can happily accept, elaborate and say: all is well, we have got a set of laws and theories which hold objectively and universally; they depict the permanent order out there. We know that the ultimate elementary particles are not so ultimate; and that they may not be particles either – in the physical sense of the word. At least on the level of quantum physics, we *process* the reality around us according to the nature of our cognitive faculties, according to the nature of our theories, according to the nature of our instruments. The process and the processor become one. The observer and the observed merge. We have also recognized that objectivity is a form of myth. The idea of 'objective' facts, and of particles which we – as it were – photograph in our theories and in our descriptions, is now seen as a glorified fiction.

All products of our understanding are necessarily shaped according to the nature of our mind, its propensities and capacities at a given time. Our mind is present in each and every theory and frame. *We always look at our mind when we look at the world.*

At one point it makes sense to say that the world and the mind are co-extensive. If mind is a part of the evolutionary process, then there is every reason to believe that future science, future forms of understanding, may be vastly different from what they are now. We constantly and inevitably co-create the world. This is the scintillatingly liberating message of the New Physics.

When our world is constantly co-created, our God cannot be frozen and static. If we accept the consequences of the New Physics – that we live in a participatory universe – then the whole idea of the fixed God and redemption theology look like a shadow of the past epoch. Only when the world is given to us, and it is brutish and nasty, and we are helpless in it, only then we must be saved from it, we must be redeemed. Since we have no powers to redeem ourselves, we have to be redeemed by God. The old God is given, static, inscrutable. We are frail and our nature is weak. The Messiah is particularly welcome in this context. Whether he will come or not, it is comforting to think that he *might*. There is no Messiah in Eco-theology, as I will argue in the course of this book (see especially Chapter 7).

The programme of seventeenth-century secularism was undoubtedly bold and far-reaching, and so were the new dreams of Reason of the eighteenth century. However, all these new designs of Secularism were only perpetuating the theology of Redemption. All the secular dreams of post-Renaissance Western society, including Marxism, and including the American Dream, are variations on the theme of redemption theology: they are mythical vessels created for our individual salvation. *Although we have claimed to have taken the responsibility for our future into our hands, we have slyly delegated this responsibility to science and technology which are supposed to be our saviours and redeemers.*

This is quite different within the New Theology, or to use my term, Eco-theology. The world is in the process of perpetual becoming. Therefore, imperfect as we are, we redeem ourselves through our own creative effort, by creating meaning within ourselves, by co-creating the universe, by creating ourselves in the image of God. *We are God in the making in the sense that the universe is in the making, partly by being received and processed in our minds.* We create realities, both social and physical, including the reality of our mind – through which we create further the world around us – either in the image of Grace or in the image of the machine. If the former is the case we live in God's universe; if the latter is the case, we live in the world in the image of the machine.

The creative act becomes all important for joining together the New Physics, Eco-theology and our individual quest for meaning. We redeem the human condition, the fact of living in the participatory universe and the idea of God evolving in us, by enacting the creative act. *At the beginning was creation.* And all life is creation. The end will come when creation will cease. Creativity is a gift. But it is also a curse. We are doomed to it – if we are to maintain our status as God's messengers.

The idea of God-in-the-making in us may sound strange or even offensive to those who are used to thinking of him as a benevolent unchanging entity benignly overlooking our vicissitudes. But to see him overlooking us is to go back to a fixed frame, is to relinquish our responsibility for ourselves, is to relegate our powers of creation onto him. When all frames are dissolved, we have changed a constant God for an evolving one. This is the consequence of a *radically* evolutionary approach. So often we have accepted evolution but only in its fragmentary manifestations. Yet nothing is excluded from its flow: it must be seen through the working of the geo-sphere, through the working of the bio-sphere, through the working of the nous-sphere, through the metamorphosis of the theo-sphere.

In short, to recognize evolution without evasion is to recognize the evolution of our comprehension, or of our understanding; it is to recognize the transient character of *all* our tools, including the tools with which we think, those very concepts and processes through which we grasp, thus understand the physical world around us; including also those ideas and concepts with which we think of God and through which we understand God. As our understanding evolves so it renders different shapes of reality, including the reality of God.

We have pointed out at the beginning of this section that as our knowledge and mind expand, God accordingly is seen in a new light, and therefore acquires new characteristics and dimensions. There are people who would wish to see God as a reality independent of our mind and of our knowledge. But even these must realize by now that God has become ecologically sensitive.

We can argue that this dimension – ecological sensitivity – has always resided in God and that we ourselves have not perceived it properly until recently. Quite so! It has been necessary for us to develop ecological consciousness in order to be able to perceive some of the new attributes of God; or should we say, to perceive some of the old attributes in a new light.

Tradition-minded people may wish to maintain that God is love, and this is sufficient to explain everything else. "God is love" (so the argument may go) entails in itself a number of other propositions, including those pertinent to the ecological interpretation of God. The proposition "God is love" includes God's love for the Planet Earth and for all its suffering creatures. God's love for the world as well as for his chosen creatures (humans) should be seen nowadays as love and care for ecological habitats, as the desire to heal the Earth, as the desire to clean our minds so that humans can see clearly and not engage in further destruction of the Earth. In destroying the planet, humans are destroying themselves and indirectly are destroying God's creation. Omniscient God would not wish his chosen creature to destroy his beloved Earth. Therefore he wishes humans to save and cherish the Earth. All of this is included in the proposition "God is love".

Yes, all ecological imperatives may have been included in the original contention "God is love". But until recently we have been simply too obtuse to perceive and read these imperatives properly. As I have said already, it required of us to develop ecological consciousness so that we could properly read and interpret God's love for the world as reverence for all life.

A true understanding of God's love for the world is expressed in our times, as our love for the Earth is expressed in ecological prayer – which is an active prayer in helping God to heal the planet, to purify the rivers, the mountains, and our own bodies. The Earth shall be fair. To make the Earth fair, we need to make our minds and consciousness clean. Cleaning our minds and acquiring ecological consciousness are the prerequisites for understanding in depth the meaning of "God's love for the world".

Summary

We do not wish to negate any previous attributes of God (Hebrew, Christian, Moslem, or Hindu). But we want to be sensitive to his recent messages. Any complete God, in our day and age, must wish us to engage in a thorough-going ecological reconstruction – if not for any other reason, then for the sole purpose of safeguarding God's creation. Ecological God simply signifies reading God's will and His imperatives in a new way, in the way that is congruent with the salvation of the planet, with the protection of other species, with the restoration of meaning in human lives, with the restoration of the meaning of God in the human universe as we approach the Third Millennium.

Our industrial, headless, materialist dance has produced the karma of pollution. This is the consequence of the reign of the rigid deterministic God. Do we have enough resolve and courage, as individuals and as a whole society, to engage in the dance of ecological catharsis, whose karma will be the life of dignity, justice and grace? Is this not what God ultimately wishes us to do?

Let us not be imprisoned by labels and old frames of reference. God is creative. That which is divine is creative. God's children – human beings – are creative. God's stupendous creation, the human mind, is creative. In the continuous process of the creative unfolding of the world, it was inevitable that we, along with the cosmic creative forces, would come to this juncture of history; that we had to start reading God's will and His message in a new way, in a constructive way. By co-creating with God, we are creating God.

The New Physics shows us how to dance in the universe in a new manner, how to dance a co-creative dance *with* the universe, how to create realities as if we were god-like. The New Physics is not yet a new theology. But its creative reach is so potent, so inspirational, so transcendent that God cannot but smile while watching how his divine universe is creatively rendered by humans.

CHAPTER SEVEN

Responsibility, Grace and Hope

Responsibility as a Force of Redemption

As we have seen from the previous chapter, the New Science (especially the New Physics) is not what it used to be – a massive deterministic bulldozer which grinds us into the small dust of insignificance as it moves in its so-called rigid, deterministic fashion. The New Physics is truly magical in its reach. Its consequences unveil to us an altogether different world.

The physicists themselves are reluctant or perhaps unable to draw larger philosophical consequences from their stories. They are especially reluctant to draw *theological* consequences from their researches. But hear it: a new view of the universe, which at the same time is a new view of the human – as we are no longer expected to be docile observers of the universe, but rather active participants in the bringing about of what potentially is there.

In this new story we cannot be passive observers, especially we cannot wait for others, or the Other, to redeem and to save us. We are our own redeemers. We are our own saviours. This is one of the consequences of the new story of the universe.

The responsibility for the fate of the universe at times appears too awesome to consider. But what other choice do we have? Must we remain docile ciphers, manipulated and disempowered by the machine and its servile experts, who tell us that the machine is too big for us to handle, that every aspect of it requires a different expert and that therefore we are to be eliminated from the decision-making-process and in fact from meaningful participation?

The new story of the universe indirectly tells us that if you understand it, you have to assume responsibility – for your own future, for the future of the planet, for the future of the universe. This last proposition is again so big and awesome that it is overwhelming. So let us express it in different terms. The universe is thinking through us and wants to take responsibility for its own fate through our wills, understanding and care – insofar as we are capable of taking responsibility for things larger than our small egos. Cultivating your own little garden in times of stress, chaos and confusion is a good strategy. But it is an escapist strategy. At this juncture of human history the universe requires more of us, namely that we become active participants in this enormous cosmic story.

Taking responsibility for things larger than your own self is nothing new in the world. It has always been a prerogative of enlightened souls. We simply need to remind ourselves that *to live as a human being is to live in the state of responsibility*. To live in the state of responsibility is the first condition of living in grace. Let us discuss the meaning of responsibility in some depth, for it is quite crucial to our new role as the custodians of the Earth and as redeemers of ourselves.

We cannot live a full human life without exercising our responsibility. *Responsibility, as a peculiar power of human will and spirit, is a crucial vehicle in maintaining our moral autonomy and in repossessing the Earth.*

Responsibility is a subtle concept. It is hard to define; and yet, paradoxically, even harder to live without. Responsibility is one of those subtle, invisible forces – like will power – for which there is no logical necessity but without which we atrophy. To reiterate, *being human is to live in a state of responsibility*. When we are unable to be responsible or voluntarily give up our responsibility, we are, in a sense, annihilating our status as human beings.

"Chosen by the gods" are those who possess a sense of responsibility bordering on obsession, like the Buddha or Jesus. "Forsaken by the gods" are those who are void of their sense of responsibility – especially for their own lives. Great spiritual leaders of humankind, as well as great social and political leaders, are stigmatized with an enhanced sense of responsibility.

The sense of responsibility is not limited to the great of the world; it is known to everybody. For what is the awareness of "the wasted life" if not the recognition that each of us is a carrier of responsibility which goes beyond the boundaries of our little egos and our daily struggles.

Responsibility, seen in the larger cosmic plan, is a late acquisition of evolution. It comes about as consciousness becomes self-consciousness, and furthermore as self-consciousness (in attempting to refine itself) takes upon itself the moral cause: the burden of responsibility for the rest. Responsibility so conceived is a form of altruism. The tendency to escape from responsibility is a purely biological impulse, a self-serving gesture, a form of egoism. Therefore, these two tendencies, the altruistic (accepting the responsibility for all) and egoistic (escaping from it into the shell of our own ego) are continually fighting each other within us. Each of us knows the agony of this fight.

When we observe the lives of great men and women, the lives that are outstanding and fulfilled, we cannot help but notice that they were invariably inspired by an enhanced sense of responsibility. Those who sacrificed themselves in the name of this responsibility did not have the sense of wasting their lives. Their examples are received as noble and inspiring.

The sense of responsibility is now built into our psychic structure as an attribute of human existence and a positive force. The negation of this force is sin, because it represents the betrayal of the great evolutionary heritage which brought us about and of which we are always aware, if only dimly.

The smallness or greatness of a person can be measured by the degree of responsibility he or she is capable of exercising for his or her own life, for the lives of others, for everything there is. Infants and the mentally ill are outside the compass of humanity precisely because they are not capable of exercising responsibility, either for their lives or for the lives of others. They are beyond good and evil, beyond sin and virtue, beyond great moral causes which propel the human family in the long run.

Though fundamental to the core of our existence, the very word "responsibility" (particularly within Protestant culture) is dreaded as a heavy burden. However, when seen as enlarging our spiritual domain, responsibility is a force that continually elevates us. "Responsibility" is a word that has wings. We must be prepared to fly on them.

We have now arrived at the context within which to view the idea of responsibility as a theological category, indeed as a pillar supporting our new religious quests. There is no doubt that we have been called upon, in our times, to assume responsibility for the future of our planet and for the future of our lives. God will help us if we help ourselves. What will finally matter is the accumulation of good Karma, good deeds performed together, rather than acts of redemption coming from heaven. *Our sin will be in failing to assume the responsibility that is thrust upon us.* Our redemption will be the act of accepting such a responsibility. We have much to learn from Eastern traditions as far as Karma is concerned. Another term for Karma is responsibility exercised.

Going Beyond Messianic Thinking

When we look at Eastern religions (Hinduism, Buddhism and Taoism) vis-à-vis Christianity, we are immediately struck that these religions are not overshadowed by the concept of the Messiah. This fact is of immense importance, but usually overlooked. In the Judaeo-Christian traditions the Messiah is always in the background. He will come to save us all. He will redeem. We wait for him. *We always wait for another to save us.* We always think that salvation, spiritual or political, is a public act to be performed by a chosen one. We want to be saved by someone – not by ourselves.

This attitude toward salvation has enormous consequences. Traditional religion may have waned; the idea of the Messiah has not. For even at the time when the idea of Church-the-redeemer (or more specifically: Jesus-the-redeemer) has lost its liturgic power, we have not stopped thinking in messianic terms, in terms of one saving the many. *In truth, we have elevated science and technology to the role of the saviour.* The tenacity with which we cling to the old idea of the saviour is truly amazing. The Messiah, of one sort or another, will come (from the outside) and save us – no matter what. This idea permeates the whole culture, although at times it is expressed in strange terms.

If you consider the idea of the "technological fix", then you become aware that it is indeed a residue of messianic thinking. And finally, take the Bomb. It is a form of radical salvation, though it may be of a perverted kind. The Bomb will clean it all, and will end all our troubles. In our murky subconscious we sometimes think of the Bomb as the "final solution". This residue of messianic thinking has grown into a pathology.

In so far as the messianic theology has developed the expectation to wait for someone to redeem us, it has indirectly cultivated irresponsibility in our midst. The whole phenomenon of modern technology, which will fix it all, is a disaster both existentially and theologically: the scope of our responsibilities has been continually shrinking while we have been waiting for the fixes. Alas, the heroin fix is an extension of the same idea! Although we don't think about technology as a God yet, we have delegated our salvation to a rather savage and insensitive God, whom I call Technos. It cannot be denied that the very idea of the Messiah is built into the structure of our present technology.

Now it may not be palatable to Christian theologians that a form of irresponsibility is built into the very notion of the Messiah. Yet we must face the issue squarely. In the concept of the saviour there is clearly lingering a notion that we ourselves cannot take responsibility for our own salvation.

It is all different in Eastern religions: no Messiah to come, no one to save you. You yourself must be your own Messiah: through good Karma, through following the path of virtuous life, through achieving Enlightenment which among other things signifies the end of suffering. When the Buddha lay dying he said: "It may be that in some of you the thought will arise, the word of the master is ended; we have a teacher no more. But it is not thus that you should regard it. The dharma (teaching) which I have given you, let that be your teacher when I am gone. Even the Buddhas do but point the way."

Yes, even the Buddhas do but point the way. Let us realize that at later times Buddhism did become the doctrine of faith at the expense of the older doctrine of merit. The devotional life of simple faith and prayer was deemed all important for salvation, not the hard spiritual path of self-perfectability. But this devotional path is characteristic only for some forms of Buddhism, while its main thrust emphasizes the path of merit, of personal responsibility, of being your own saviour.

Waiting for *the other* to save you is a denial of your authenticity and your responsibility. Working on yourself is an expression of the trust in your self-perfectability. Those who have worked on their spiritual path can not only help themselves, but also emanate the subtle radiance (sometimes called grace) which helps others.

We shall not deny that the path of spiritual self-development has been known within the Christian tradition. Yet it has not become the mainstream Christianity. Also, it has not been sufficiently realized that the idea of the saviour who will save us all is dis-empowering in the long run; it contains the licence for apathy and, indeed, for irresponsibility.

Yet there are forms of responsibility which lead to pathologies. Hitler exercised his peculiar sense of responsibility when he at-tempted to establish the supremacy of the Aryan race at the expense of other races. *It is clear that a sense of responsibility is not enough in itself. It must go hand in hand with a right eschatology, a larger purpose on behalf of which this responsibility is exercised.*

When the sense of responsibility is wedded to and inspired by a wrong eschatology, or by a perverted sense of ultimate purpose, tragedies may follow as happened in Hitler's Germany or in Stalin's Russia. Therefore, what is needed is a universal consciousness or the Great Compassion, as Buddhists call it: a sense of unity with all, a feeling of empathy for all, in short, a framework of universal altruism, a right kind of spirituality. When the framework is in-formed by compassion and inspired by the sense of co-working with evolution, then the sense of individual responsibility will not breed Stalins or Hitlers.

Let us be aware that responsibility which goes beyond one's selfish concerns cannot be justified rationally. There is no *logical* reason why one should act responsibly on behalf of all humanity and of evolution. Ultimately, responsibility is a concept that belongs to the religious realm. You are responsible for the world because you care, because ultimately you deem the universe to be sacred, thus worthy of passionate care.

Grace: The Result of Karma

What is the place of Grace within Eco-theology? Grace does not come to those who do not invite it, who are not ready for it, who do not work for it. Yet oftentimes it does not come to those who work for it and invite it. In this sense grace is a gift. But is it a supernatural gift or a part of our natural endowment? Christian theology will insist that it is the former; Eco-theology maintains that it is the latter.

In the traditional Christian framework grace is explained *deus ex machina*, by the intervention of divine powers residing outside ourselves. Participation in God is then an act of Divine Mercy bestowed upon us from outside.

We need to look at grace *de novo* and see it in a new light. This new light will inform us that grace is not supernatural but, on the contrary, a part of our natural condition. Grace is a consequence of our status as human beings. *To be divine is a natural condition of being human.*

Though natural, grace does not come easily as breathing for instance, for grace is releasing the God within, and this process is subtle and painstaking. One of the methodologies for reaching grace is meditation. But meditation is a tool, not an assured road, not grace itself. We should not mistake the means for the ends. Grace may come independently of meditation but it usually comes in the condition of inner silence.

Is a gift of grace more special than a gift of music? At first we are inclined to say yes. Upon deeper reflection the answer is not so obvious. Was Mozart not in a state of grace when he composed his celestial pieces? And why do we call them "celestial"? Clearly because at one point a great musical talent is a form of grace. Yet we are not inclined to call a musical talent a supernatural gift; rather a part of the natural human endowment. We are all gifted musically – to a degree. Some are gifted to an extraordinary degree.

Might it not be reasonable to suggest that we are all endowed with, at least, some rudiments of grace? Some are endowed more, some are endowed less. And some indeed work extraordinarily hard to bring out the latent grace in them to full blossom. *Grace, like a great musical talent, has to be nurtured, trained and developed in order to come to fruition.* The story of grace as revelation omits the element of endless toil, the great discipline and the countless exercises that the illustrious ones had gone through before they achieved the state of grace. The Buddha wandered for three years, fasting nearly to death, before he was visited by Enlightenment and then lived in the state of grace. Jesus "disappeared" for eighteen years – possibly spending most of this time in an Essene community. Doing what? We don't know. Probably not carpentry; but rather, strenuous spiritual exercises.

In short, great spiritual leaders were supreme because they exerted their will power to a supreme degree while eliciting their potential for grace. We ordinary human beings so often fail not because we do not have this potential within us, but because we do not have the will power, the fantastic discipline, the inner resilience to keep going – no matter what obstacles ordinary reality puts in our path.

In this sense grace is the result of good Karma, or right Karma, the fruit of one's inner workings; a part of the natural endowment brought to an extraordinary fruition.

There is no royal road to grace, nor any set of prescribed, reliable exercises. Our spiritual life is too subtle and too complex to produce grace as the result of some specific method. However we know that the chosen ones invariably went through rigorous spiritual practices. So the inner discipline and spiritual practices are a necessary condition, but not a sufficient one.

This condition of wholeness, which we call grace, invariably goes beyond the wholeness of one's individual being, beyond the integration within the human person. For grace radiates harmony which is of a cosmic kind. For this reason we imagine its source to be in heaven, or simply a special gift of God. One of the characteristics of grace is that it is a form of love. Grace is also a form of radiance which, like hope, enables others within its reach to be more dignified, more human, closer to their inner selves.

Grace is the flower. In its radiance are reflected all other attainments of the human heart and spirit.

We are all children of God, born with divine attributes. There is no reason to assume that Indian Yogis and Buddhist saints are precluded from the life of grace because they do not subscribe to the Christian creed; indeed, many of their lives are exemplars of living grace: the life of Gandhi, for example. One of the divine attributes of the children of God is the capacity to reason. Another is the capacity to love. Yet another is the capacity to bring grace to fruition in their own lives. The will of God reveals itself but slowly.

Hope and Courage as the Vehicles of the New Theology

Eco-theology joins hands with Paul Tillich (1886–1965) in celebrating hope and courage as both attributes of our humanness and characteristics of the New Theology which envisages the human being as God-in-the-making.

The ethical question concerning the nature of courage, and the ontological question concerning the nature of being, are intimately and inextricably woven together. Thus writes Tillich:

> … an understanding of courage presupposes an understanding of man and of his world, its structures and values. Only he who knows this knows what to affirm and what to negate. The ethical question of the nature of courage leads inescapably to the ontological question of the nature of being. And the procedure can be reversed. The ontological question of the nature of being can be asked as the ethical question of the nature of courage. Courage can show us what being is, and being can show us what courage is.
>
> *The Courage to Be,* p. 2

The question of courage is by no means theoretical or theological. Western scholars live nowadays in comfort and freedom, with their jobs well secured, and with their pensions well provided for. Yet courage is totally lacking in Western universities. "Intellectual" courage is not courage but hiding behind the screen of established orthodoxies. The roots of real courage are invariably moral. Present academics, by and large, have no firm moral convictions. Living by the creed of rationality leads to the atrophy of courage in a deeper sense of the term. As they do not have it themselves, they cannot instil courage into their students. Thus we are breeding, in the universities, a spineless generation, with no courage and no moral sense; the two are connected, and the consequences are devastating.

The courage to be is for Tillich one of the most fundamental attributes of human existence. Yet this courage needs an evolutionary dimension. A unique feature of our humanness in us is not only the courage to be but *the courage to become*, the courage to continually transcend every station we have reached. The courage to become is one of the defining characteristics of Eco-theology, and of the Ecological God.

An understanding of this courage to transcend gives us a unique perspective on the nature of humans and their values. Only he who knows that to be a human being means not only the courage to be but also the courage to build beyond and the courage to leave behind every station we have arrived at, will understand the agony and beauty of past strivings either of individuals or societies or species, as well as present frustrations and despairs that we are not able to do more than we can actually do. We rationally understand our limitations but in our deeper instinct, in our transcendent vision, we know that we can and must overcome these limitations as our destiny is nothing short of divine.

The right understanding of the courage to be and the courage to become leads us to the ecology of hope. Hope is the foundation of our courage – the hope eternal in spite of all the vicissitudes of our fortune. The restoration of hope is particularly important in our times, eaten by scepticism and nihilism. In restoring hope we demonstrate responsibility in action. The ecology of hope must be tenderly cultivated amidst the ecology of destruction. Indeed it looks as though the destructive demon has created its own ecology, which is a downward spiral sucking us into the centre of the abyss.

Blossoming of hope is a prelude to grace. Withering of hope is a prelude to death. How can people live who do not live in hope? Hope is the spring that renews us daily. Hope is as fundamental as the oxygen we breathe. Hope is the scaffolding of our very being. Hope is the ray of light that separates life from death. We must burn the candle of hope like an act of self-sacrifice.

Like Lazarus, we must rise from our broken condition to embrace the new dawn of hope which is the light of our human condition, and without which our human condition is steeped in darkness.

Hope and faith are ontological underpinnings of our being. Our rational mind seems to have forgotten that faith and hope are not foolish indulgences of "weak" people who are not tough enough to confront the world in a tough way, but the very roots that nurture our being.

In Peter Shaffer's play *Equus* we are confronted with the line: "Without worship man shrinks".

The ecology of hope goes a step further as it maintains that if you worship nothing, you are nothing. To believe is natural for human beings; as well as to worship something. Faith, hope and reverence are therefore the necessary ingredients of our well being, they are integral dimensions of our wholeness.

In brief, hope is a mode of our being. A healthy human being cannot live outside the gentle mantle of hope. Hope is a pre-condition of our mental health. It is also the awareness that human destiny is not determined by blind destructive forces that will crush us no matter what. Hope is a reassertion of our belief in the meaning of human life, and in the sense of the universe. Most importantly, hope is a pre-condition of all meaning, of all strivings, of feeling at home in this confused and complex world of ours. Thus endorsing hope is not a form of foolishness but a form of wisdom.

Teilhard's Predicament

The history of the Western Christian Church has been nothing short of turbulent. The last century has seen the Church at its lowest ebb. The time of crisis is also the time of reformers who attempt to put things aright. Of the many important Christian theologians of the twentieth century the following three should be mentioned: Schweitzer, Teilhard and Tillich. Schweitzer (at least at his best) can be easily accommodated within Eco-theology. The heritage of Tillich has been discussed. In this section we shall discuss the ambiguous theological heritage of Pierre Teilhard de Chardin (1881–1955).

There are two opposite forces within Teilhard's theology which are never satisfactorily reconciled in his own writings. Teilhard's predicament is also Schweitzer's predicament; it is really the predicament of twentieth-century Christianity: how to reconcile the Christian teaching based on dogma, revelation, the idea of salvation, as well as the idea of the return to Paradise with the new insights, a new *Tao* that accentuates the process of self-perfectability, of responsibility, of the idea of man/woman who co-create with God? In more explicit terms, we witness a clash between the past dogma and the longing to actualize within ourselves the Cosmic Christ; the clash between the authority imposed on us, and the sense of our own responsibility for our own life as well as for the life of the globe.

If we take Teilhard's chief opus *The Phenomenon of Man*, it is clear that his God is evolving towards the Omega Point; and only then, at Omega, will he become Full God. Such is the main thrust of the book. However, at the very end of the book, in the epilogue, Teilhard changes the whole perspective and, as it were, tries to "mend" his way. He explicitly acknowledges that his Omega God is in the image of the traditional Christian God. This epilogue is a strange thing. After he has developed a new theological perspective, Teilhard "catches" himself in heresy, becomes aware of his schism, and proposes to return to Christianity. It thus appears that Teilhard forced himself to reinterpret his creative evolution in the Christian key as he writes:

The universe fulfilling itself in a synthesis of centres is in perfect conformity with the laws of union. God, the Centre of centres. In the final vision the Christian dogma culminates. (sic!) And so exactly, so perfectly does this coincide with the Omega Point that doubtless I should never have ventured to envisage the latter or formulate the hypothesis rationally if, in my consciousness as a believer, I had not found not only its speculative model but also its living reality.

The Phenomenon of Man, p. 322

Teilhard tries to equate the process of evolutionary unfolding with the Christo-Genesis. But the equation does not work because it cannot work. The two processes of conceiving of God – the traditional Christian and evolutionary – are *diametrically* opposed to each other. By attempting to subsume Omega Point under Christian theology, Teilhard undermines the *raison d'être* of evolution as an unfolding and self-actualizing process. For if evolution is a return to Christ, a return to the original Paradise, then *it only recapitulates the past*. If, on the other hand, evolution is actualizing itself, and will be only actualized at the end of time, then *there is no return to the Paradise Lost, for there never was a Paradise before*.

Let me discuss this dilemma in some detail, for it is of great importance. If we recognize the notion and the authority of God as conceived in the traditional religions, particularly in traditional Christianity, then our evolution, including the moral and spiritual one, is completed. What we can do, and the *only* thing we can do, is to return to the Paradise Lost, to re-acquire virtues that have been bestowed upon us by God-the-Original-Maker.

If, on the other hand, we see ourselves as unfinished spiritual beings, indeed only in the infancy of our evolution, then we simply cannot accept the notion of the traditional God *who made us perfect*. When we contemplate our primordial beginnings in the cosmic dust, we realize that they were far from godly. It is only at the end of the Immense Journey that we may become godly; but only if we actualize God in ourselves. God is in the making – in us. The further we go in our evolutionary journey, the closer we may approach him. God is spiritually actualizing himself in us. Our goal should therefore be to transcend further and further and never to return. For a return represents a fall from grace. Evolution is a barbed wire. If we could return, it would be a return to brutishness and dimness and not to grace and perfection. In so far as Teilhard upholds the Christian myth of Return, he profoundly undermines his evolutionary thesis. In so far as he seriously upholds the evolutionary thesis, he profoundly undermines the Christian heritage. There is thus a profound incoherence in his views.

But there is a way of incorporating the Christo-Genesis into the evolutionary design, namely by treating Jesus not as God, a point of final destiny and of ultimate strivings, but as a symbol, as an inspiration, as a reminder that even at this stage in our evolutionary development we *are* capable of so much grace and divinity. Then Christ-consciousness becomes not so much the ritualistic identification with Christ's body or blood but an imaginary flame that illumines our roads towards greater grace and consciousness.

Another path to reconciliation is to think of the whole universe in a sacramental way – a view not entirely foreign to early Christianity but not one that is actively entertained and pursued by present Churches.

Our culture is woven around the skeleton of Christianity. To deny or destroy this skeleton would be nothing short of destroying Western culture. Cultures live by symbols. Symbols of our culture have been thoroughly pervaded by Christian teaching and metaphors. We cannot tear apart the vessel in which we are floating. Like Descartes, we have to reconstruct our boat while we are afloat.

Our task is made even more difficult by such people as Lynn White, who has insisted that the blueprint for the destruction of the environment is in the Bible. In his celebrated essay, "The Historical Roots of our Ecological Crisis", White has galvanized our attention, and almost paralysed us with inaction, by telling us that in a sense we are doomed. The most important document of our culture, by which we have lived, is fundamentally misconceived. As a guidance and inspiration it has led us astray. White has eloquently argued that the Judaeo-Christian theology is responsible for the separation of man from nature. "Man's relation to the soil was profoundly changed. Formerly man had been part of nature; now he was the exploiter of nature.... Man and nature are two things, and man is master."

Here, therefore, was the evidence showing that the sacred book was, if only implicitly, a blueprint for destruction. During the last twenty-five years we have absorbed the shock of this "revelation" and slowly have learnt to live with it. We have come to recognize that our culture is not made of one cloth but possesses various layers and is open to different interpretations. What we have witnessed during the last decade is a conscious effort, within both the Judaic and Christian traditions, to reinterpret our religious and cultural heritage, so that we can make peace with nature, and *ipso facto* with an important part of ourselves.

Summary

To be a human being is to live in a state of responsibility. The sense of responsibility is built into our psychic structure. The smallness or greatness of a person can be measured by the degree of responsibility he/she is capable of exercising. Our sin is failing to assume the responsibility that beckons us. Our redemption is an act of assuming such a responsibility.

The messianic thinking of Western culture has led to the disempowerment of individuals. With a new mandate of responsibility thrust upon us by the participatory universe, we are destined to be our own messiahs. We cannot wait for Godot forever, particularly in our urgent times when God wishes us to take responsibility of his creation and thus be our own saviours.

Grace is a condition of our being, of our wholeness. Grace is a form of radiance. There is no royal road to grace. Those who have attained it work incredibly hard on themselves. The potential for grace lies in each of us. But to bring this potential to blossoming requires a towering act of will-power.

In all our aspirations – earthly and heavenly – we have to hold to hope tenaciously and resolutely. The blossoming of hope is a prelude to grace. Withering of hope is a prelude to death. The way to heaven is paved with hope. The way to hell is littered with hope abandoned.

"Adore and what you adore attempt to become" (Aurobindo).
"In truth who knows God becomes God" (Mundaka Upanishad).
He who affirms the divine, affirms himself.
The divine kingdom is yours if you dare to enter it.
Link yourself with the highest light and become it.

Those are the statements projecting our being into what we are not yet. They should not be treated literally, but rather as guide-posts helping our inner journey. In order to be, we have to enlarge our being continually, we have to have the courage of becoming. In order to enlarge our being, we must have a place to go – not in the trivial economic sense, but in a cosmic sense.

At this juncture of cosmic history, before higher evolutionary forms emerge and supersede us as more efficient vehicles of God-making, we are God-in-the-making. In the minimal terms outlined here is a methodological programme: how to make more of the universe that is nurturing God, how to make more of ourselves while continually fusing meaning into our lives.

CHAPTER EIGHT

The Challenge of St Francis

The Franciscan Path

Christianity has been in crisis for a number of centuries – at least since Luther. Its acute crisis started in the industrial age, and has accelerated in the technological age. Faced with the cornucopia of the technological abundance which promises salvation through consumption, the Christian churches have shrunk in their reach and responsibility. *They have been unable to meet the challenge of growing materialism* which emerged in the wake of the triumphant technological world view – which itself declared the world to be a machine, manipulated by technological devices to man's advantage.

Thus, the churches have been unable to face the consequences of the scientific-technological world view which offers a total interpretation of the world without any role of religion in it. On another level, the churches have been unable to meet the moral challenges of cynicism, nihilism and the value vacuum which have crept in as the companions of material affluence. The churches have been unable to offer an inspiring moral and spiritual challenge which would combat the growing cynicism mainly because Christian spirituality has been burning low in recent decades if not centuries. A church, any church for this matter, cannot infuse a radiant spiritual inspiration into its followers if its spiritual energy burns low or is exhausted.

The overall malaise of the Christian churches has been deepened by the ecological crisis. The simple truth is that Christian teaching is ecologically insensitive. Moreover, some have found the Bible to be a blueprint for ecological destruction. I am referring to Lynn White's essay again. There have been many rebuttals to this essay, mainly written by Christian theologians and clergymen. These rebuttals help but little when the Bible can still be seen as ecologically insensitive, and moreover when we realize that, within the body of Christian teaching, the Franciscan perspective has been so profoundly neglected.

In this chapter I will show that, although neglected, St Francis has always been a source of sustaining power within Christianity, a good shepherd who makes us aware of the power of simplicity. I will attempt to show that the Franciscan teaching has survived much better than the teaching of orthodox Christianity simply because St Francis' vision is deeper and more universal than the one officially followed by the church. A return to the Franciscan vision may be a salvation for the Christian churches.

Let me first briefly outline Francis' life and thought. He was born in Assisi in 1182, died in 1226. Thus he lived for only forty-four years. As a young man Francis was ambitious. He dreamed of fame and honour but without any definite idea how fame was to come to him. After his sorry participation in the war with Perugia, Francis did not give up the idea of the earthly glory to be won with a sword. He enlisted as a voluntary soldier for the war that was being fought at the time between the Papal armies and the despised German armies, on the islands of Sicily in the years 1200–1202. While on the way to the battlefield, he heard strange voices during his sleep, calling him to serve the Lord. He heard the same voice the next day, this time being only half-asleep. The voice urged him to return "to the land of his birth". And so he did. This was the end of his military career, and the beginning of his spiritual quest.

After months of lonely struggles with his soul, while continually praying and often weeping at the Etruscan tombs outside the town of Assisi, he spent a day at St Peter's church in Rome as a beggar in beggar's clothes with his hand outstretched for alms. Through this experience he found a kinship with the poor. He embraced Lady Poverty.

Now, on the slope of the hill outside the city of Assisi, there stood the little church of San Damiano. Francis took a liking to it. The church was in a sorry state of disrepair. One day while praying there, he heard a voice, "Francis, repair my church." He returned to his father's house, packed a big load of finest cloth, loaded it on the finest horse his father possessed, and went to the market. He sold the cloth and the horse, and brought a considerable amount of money to the priest of San Damiano. The poor vicar was too frightened to accept the money after he heard the whole story. So Francis left an urn with the money on a window sill.

Upon returning to Assisi after his journey, Francis' father Pietro Bernardone learned about his son's "theft". He got angry, then furious. He decided to try Francis before the magistrate of Assisi. Francis refused the summons, declaring that being a man dedicated to religion, he was not subject to the civic authorities but only to the bishop.

Thus, Bishop Guido was to preside over the trial of Francis in the cathedral of Assisi. On the day of the trial, Francis duly appeared. When remonstrated by the bishop to return the money to his father, Francis said: "My lord, I will gladly give back to him not only his money, but all my clothes I have had of him." Then he took off all the clothes he wore. Stark naked, he put his clothes in front of the bishop.

This act required more than courage. This was an extraordinary act of confronting wealth and authority with one's naked body in the middle of a cathedral. This was an act declaring: Here I am, at God's tribunal, and may frail human justice tremble in the face of deeds done on behalf of the needy and the oppressed – even if they violate the existing legal codes.

The rest of Francis' life was a continuation of the dramatic encounter at Assisi cathedral. The essence of the Franciscan legacy can be expressed simply:

- Confront injustice and human misery directly. He who does not speak on the behalf of the oppressed, contributes to the crime.
- Redistribution of wealth is our responsibility. The unjustified accumulation of wealth by some is the root cause of miseries of others.
- We stand before God's tribunal, which ultimately means the tribunal of our conscience. To this tribunal of our conscience we are ultimately responsible, particularly as the existing laws so often protect and favour the rich at the expense of the poor.
- Have the courage of being naked. Ultimately you are naked. No garment should muffle the voice of your conscience.

- Have the courage of simplicity. For this simplicity can release from within you great spiritual powers which are numbed by your slavery to mindless over-consumption.

There is no question that Francis was in advance of his age, as he anticipated all that is liberal and sympathetic in modern times: the love of nature, the love of animals, the sense of social compassion, the sense of the spiritual dangers of affluence. Of various aspects of his important legacy, the most important for us is Franciscan ecological or ecological-spiritual legacy. The worship of nature was for Francis a part of the overall ecological spirituality. We thus hear these incantations expressed in *The Canticle of Brother Sun*:

> Be praised, my Lord, through Sister Water,
>   who is very useful and humble and precious and pure.
>
> Be praised, my Lord, through Sister Moon and the Stars,
>   in the heavens you formed them clear and precious and beautiful.
>
> Be praised, my Lord, through Brother Wind
>   and through Air and Cloud and fair and all Weather,
>   by which you nourish all that you have made.
>
> Be praised, my Lord, through Brother Fire,
>   by whom you light up the night;
>   he is beautiful and merry and vigorous and strong.
>
> Be praised, my Lord, with all your creatures,
>   especially Sir Brother Sun,
>   who is day and by him you shed light upon us.
>   He is beautiful and radiant with great splendour,
>   of you, Most High, he bears the likeness.
>
>   *Writing of St Francis of Assisi*

Because of his extraordinary sense of empathy with all creation, because of his gaiety, romantic imagination and universal camaraderie, not only Christians but people of all religions who are acquainted with his teachings are drawn to Francis.

St Francis' discourses were not numerous, not elaborate, let alone learned. He did not produce anything comparable to the *Summa Theologica* of Thomas Aquinas. He did not produce any systematic theology. His was the vision of the heart, speaking directly to the heart of others. Of his writing, perhaps one of the most touching is one of his prayers:

> Lord, make me an instrument of Your peace.
> Where there is hatred, let me sow love.
> Where there is injury, pardon,
> Where there is doubt, faith,
> Where there is despair, hope,
> Where there is darkness, light
> And where there is sadness, joy.
> O Divine Master, grant that I may not so much seek
> To be consoled, as to console;
> To be understood, as to understand;
> To be loved, as to love;
> For it is in giving that we receive,
> It is in pardoning that we are pardoned,
> And it is in dying that we are born to eternal life.

Francis was not the first man to make voluntary poverty an enormous shining virtue. The Buddha embraced the same ideal 1,800 years before Francis. He too renounced all the riches of earthly splendour; abandoned his palace, his wife, his son, and in beggar's clothes went to seek salvation and redemption. In our century, Mahatma Gandhi, inspired by Hindu ideals, followed a path of simplicity very similar to that of Francis. Those saints of other religions do not diminish Francis. On the contrary, in their company Francis shines as a universal being, as *a cosmic being*. For the ideals he proclaimed ring true in all major spiritual traditions of humanity. *Francis is the saint for all religions*. Among Christian saints, Francis is the most popular and most liked. There is something deeply appealing to the human heart in the idea of renunciation, in the idea of stripping one's self of unnecessary superficial layers, in the idea of going naked to commune with nature and all its creatures.

Christianity Fails to Embrace Franciscan Teaching

It was a misfortune for Western culture and Western Christianity that it chose the path of Thomas Aquinas and not the path of St Francis. From the beginning, St Francis was a challenge extraordinary to the

Roman Church, and an extraordinary inspiration to simple Christians. He was on the verge of being excommunicated more than once. He was canonized very soon after his death – in order to avoid a split within the Church. To coopt him as a saint was a way of blunting the poignancy of his message – that the life of a Christian should be one following Christ's simplicity.

For a couple of centuries after his death, the Franciscan way was alive and vibrant, if only among ordinary people. Then came Martin Luther with his challenge. And then came the Council of Trent of 1545–1563, which tried to find a radical and satisfactory response to Luther. Since the Trent Council, Thomist Christianity has prevailed, and the Franciscan vision has been pushed aside. Why? The reason is not a direct one. It was not the case that the Bishops of Trent loved Aquinas more than St Francis. St Thomas was a dry pedantic scholar, and not a lovable warm human being.

The great problem of the time was how to contain Luther. There were many tricky and muddled theological problems, and Aquinas was so clear in spelling out the Christian doctrines. His vision of Christianity won.

The power to define is the power to control. Thomist Christianity is based on the power of the word, which is the *power of the mind*. Franciscan Christianity, on the other hand, is based on the power of love, which is the power of the *heart*. Within the Christian tradition, the mind has prevailed over the heart. And Christianity has become brainy, abstract, wordy – and void of heart. This is our dilemma nowadays, of the whole Western civilization: we are clever and brainy. But our hearts have atrophied. The Thomist version of Christianity has undoubtedly contributed to this dryness of the heart as it has promoted the verbal, abstract, brainy tendencies of the mind.

Whether the Council of Trent was a decisive point in crystallizing Christianity around the Thomist doctrines is hard to ascertain with any definiteness. The momentum in this direction was already apparent in the Middle Ages. In the fourteenth and fifteenth centuries the theological disputations over the nature of God became exceedingly abstract. The way of the mind was clearly suppressing the way of the heart.

Now St Augustine and the early fathers of the Church were very comfortable with Plato. The mystical approaches to Christianity were flourishing until the time of St Thomas Aquinas. However, as time passed, the rational attitude began to prevail. Aristotle replaced

Plato as the foundation of Christianity. Thomist theology (we must remember) is Aristotelian through and through.

Thus Christianity repeats the road (perhaps we should say *the mistake*) of ancient Greek philosophy. As we remember, early Greek philosophies, including that of Socrates and Plato, revelled in holism and bathed in the all-pervading sense of mystery surrounding us. With Aristotle we begin the quest for analytical clarity. For the sake of this clarity, things are cut up and parcelled off. Rational justification and analytical definitions are arrived at at the expense of depth, mystery, wholeness and this ineffable feeling of oneness with things which pervades the universe; that is, if one believes in the connectedness of things and the mystic union of the Cosmos. Thus Christianity, under the guidance of Thomas, repeats and re-articulates the road taken by Aristotle, a great mind, but a tedious pedant among ancient Greek philosophers.

We should be aware that in the early Middle Ages, mystical interpretations of Christianity abound. The twelfth and thirteenth-century mystics have left quite an inspiring and important legacy. They were often suppressed by orthodox Christianity, and their teaching not infrequently distorted – to mention nothing of the fact that many of these mystics were excommunicated and officially condemned by the proclamations of the Vatican.

St Francis was one of the mystics. He recognized that the power of the heart is greater than the power of the mind, that the power of things unseen is greater than that of things seen, that the power of simplicity, of frugality, of giving, is more significant than the capacity for acquiring and luxuriating in earthly splendour.

Franciscan powers are exactly those preached by Jesus Christ. Thus Francis will always remain relevant, appealing and inspiring to those who take the teaching of Jesus seriously. Francis will continually remain a challenge, if not a thorn, to the institution of the Church, particularly when this institution insists on the primacy of the doctrine over the understanding of the heart.

The attitude towards mysticism in the Western world is thoroughly ambiguous. We are uneasy about mystics. They are just not "rational" enough. We want to keep them at arm's length – as far as our rational mind is concerned. *Our whole culture has been sanitized by rationality.* Anything holistic, deep, mysterious, let alone mystical, is often condemned as irrational. We need to remind ourselves again of what Albert Einstein says about the mysterious:

"The most beautiful thing we can experience is the mysterious."

History is a great illuminator, and a great tyrant. An illuminator, for it explains what has happened and why. It is a tyrant because it makes us believe that what has happened was inevitable. Things could have happened otherwise – in so many ways. We know and feel in our bones that Christianity could have followed quite a different path. It was a great misfortune for Western culture that Christianity did not follow the Franciscan path – which was such a distinctive possibility.

Yet looking further back into the history of Western civilization, one has the feeling that from the sixth Century BC onwards, the dice had been loaded. Once the Greek logos had emerged and crystallized itself, the Western mind was in a sense bound to follow its "rational" path.

Yet we need to be aware that the logos of Pythagoras and of Plato was different from the logos of Aristotle. Before Aristotle's time, the logos was all encompassing, was in fact a total divine matrix for deciphering the meaning of the cosmic laws. The divine was as much a part of this logos as was the mathematical and the rational. Thus we should be aware that for the Pythagoreans mathematics was suffused in divinity.

It all changes with Aristotle. We come back to Aristotle over and over again because he is a dividing line. With Aristotle we open up a glorious period of Western rationality, of fine analytical scrutiny which leads us to fantastically powerful explorations of the Cosmos. With Aristotle, alas, we lose the sense of wholeness, and in the process *we lose a sense of the divine*. We have found of late that rationality is a poor substitute for spirituality.

The great question that confronts us is: was the Western mind so made that it had to develop its discursive logos with its distinctive, hard-cutting rationality? Or is this rationality the result of a fortuitous development, the result of the fact that we followed Aristotle and not Plato? And later that we followed Thomas Aquinas and not St Francis? I believe that the latter is the case.

Whom we follow has enormous consequences for our culture and for our lives. When we find that our culture is breaking up and our lives are messy and incoherent, then we go to the sources and blame the prophets who have led us astray. We are now at such a point of history.

Let us reflect, therefore, what would have happened to Christianity and Western culture had we followed the Franciscan path. This kind of meditation on the past is not an exercise in futile nostalgia, but an exercise in visionary imagination which may help us to see more clearly our present and future. I'll claim in the course of this essay that the Franciscan path is still open to us.

Had we followed the Franciscan way, the devastations of nature would not have occurred, and the devastations of our individual lives would not have happened – the two are aspects of the same phenomenon. The Franciscan teaching that all beings in creation are our brothers and sisters is a very powerful and inspiring guide of how to treat nature. This is what the Native American Indian cultures have believed and followed to the benefit of ecological systems for millennia, until they were ravaged by European diseases and settlers. It may be said that Native Americans were good Franciscans; or that St Francis was a good Native American.

How we treat nature is ultimately a religious matter. How we treat animals is ultimately a religious matter. How we treat each other is ultimately a religious matter.

The way of the heart, within which all creatures are sacred and are our brothers and sisters in creation, would have enabled Christianity to extend justice to all beings in the universe. The Franciscan way would have precluded the destruction of nature from the onset, for nature is not a thing "out there" given to us for our use, but a part of our outer self. The Franciscan way would have brought with it a different form of perception, a different form of thinking (which I call reverential thinking) and consequently different paradigms of what is good and effective action.

St Francis, Our Contemporary

The enduring beauty of a work of art, or of an extraordinary idea, lies in its capacity to shine through centuries after its origin. St Francis' ideals are of such a nature. There is a powerful ring in his ideals of simplicity and of the communion with all beings which is deeply felt, not only among Christians, but also among people of other religions. There is thus enduring universality in Franciscan ideals and visions. They seem so simple that we are inclined to think that they are too simplistic. Yet this simplicity is the source of their enduring inspiration and their enduring power.

In 1979, Pope John Paul II proclaimed St Francis as the Patron Saint of Ecologists. Which was very nice. But more a gesture than the beginning of a substantive change in the policy and teaching of the Church. As I have already mentioned, Francis was canonized very soon after his death. One suspects a similar thing happened when Francis was proclaimed the patron saint of ecologists. The Vatican seems to be saying: "You have now got St Francis as your patron. You should regard this as an indication that the Church cares for ecology." However, in order to make Francis a part of the living Christian liturgy, and a part of the actual lives of actual Christians, much more is needed than making St Francis a patron saint of ecologists.

For the main problem still remains. This is the problem of elevating the human person above all other creatures, and giving man the dominion over every living thing that creepeth on this earth. This is the traditional teaching of the Christian church, *which has been ingrained upon our consciousness*. If we are to acquire ecological consciousness we need to alter fundamentally the nature of this teaching.

According to Christian ethics, human life is sacred because we are the children of God, because God made our lives sacred. From the sacredness of life, bestowed on us by God, follows human dignity, which is an aspect of this sacredness. This design possesses a great strength. In it, human dignity follows automatically from the status of man as a privileged creature of God. But this conception also presents severe drawbacks. The favourite status is reserved for man alone. He is God's chosen creature. His life is sacred, and therefore must be treated with dignity. Other creatures, and nature itself, do not possess this status. In brief, *Christian ethics is ecologically mute*.

The truth is that religious ethics based on the Bible, whether of the Jewish, Christian or Muslim variety, have little power in restraining the devastating impact of the chariot of secularism. *We need much more than the conviction of the sacredness of the human person*. Unless this sacredness is extended to other creatures, the gulf will always remain between us and the rest of creation.

Many people who subscribe to Christian ethics, on a personal level, are at a loss how to use it to heal the wounded planet. There is no easy way to translate the ethics based on the Ten Commandments into meaningful ecological action *as long as man is on the pedestal*, the sacred being, while all other beings are down there, not accorded a reverential status. The reinterpretation of the traditional doctrines of the Church, including ethical doctrines along ecological lines, will be a major task for which the Church does not yet seem to be ready. In these reinterpretations St Francis should be our guide. *For he represents a new promise for Christianity*. It was not only possible for him but indeed *inevitable* to celebrate the sacredness of man *and* the sacredness of all other beings.

We need to return to the teaching of St Francis in earnest. From his visions and views concerning the oneness and sacredness of all creation we can derive new, ecologically sound Christian ethics. *A new encyclical should be announced by the Pope proclaiming that all creatures are sacred*. We should all work towards this end – by making the Church and the Pope aware that unless all nature and its creatures are declared sacred, we do not have a solid ethical basis for a radical healing of the planet and for eliminating environmental pollution – which is now a more serious threat to the future of human kind than the possibility of a nuclear holocaust. If we lose the environment, we lose God.

The Importance of Social Justice

Marxism is dead, and Communism has collapsed. Such are the pronouncements of our times. Indeed, one conception of Marxism and one specific form of society based on it has exhausted its creative potential and now trails towards the exit of history.

Some have greeted the collapse of Marxism with a cry of triumph. For many, the collapse of Marxism signifies the undoubted triumph of capitalism. The threat of treacherous Communism has abated. We are again in the firm hands of the Almighty Free Market.

But all is not well. The triumph of capitalism is but an illusion. And the departure of socialism (by which I mean a social doctrine which cares for social justice, particularly for the underprivileged) is only temporary. We need a new theory of social justice which will be strong enough to protect the poor and the exploited, wherever they are. We need a new form of Marxism, even if we don't call it Marxism. We need what I would call *Franciscan Marxism*, a social doctrine based on spiritual foundations inspired by the ideals and visions of St Francis and the magnificence of the miracle of life.

Many people have watched with alarm the direction of change in Eastern Europe after the collapse of Communism there. That the Marxist regime in those countries was stupid, oppressive and dehumanizing there is no question. Its departure was welcome. But let us not throw out the baby with the bath water. The ideals of social justice, free education for all, and free access to public health have been undoubted achievements for the underprivileged classes in Communist societies. It is indeed alarming to see how quickly ex-Communist countries, such as Poland, Hungary and Czechoslovakia, have uncritically embraced capitalism lock, stock and barrel – as if it were the royal road to salvation.

For all is not well in the citadels of capitalism itself. Capitalism has not learned. Whenever it has a chance, it reveals its greedy, ruthless and exploitative tentacles. During the last ten years, the rich countries have become richer and poor countries have become poorer. We all learn these facts and statistics, and somehow accept them as the decrees of God. We hardly ever stop to ponder that these facts spell out an increasing injustice, inequity, deprivation of human dignity, and yes, in the long run, a potential for violence.

In the most prosperous of all countries, the USA, the inequities have grown. It was fashionable in the mid-1980s to maintain that *everybody* can become rich. Those who are poor *do not want to help themselves*. Yet, after ten years of pursuing this pseudo-philosophy, a clear picture has emerged. The rich have become much richer but at the expense of the poor and the middle class. The number of homeless has dramatically risen in the USA in the last ten years and now is counted in millions. The middle class has been squeezed and short–changed. In order to keep up their standard of living people have been working much longer hours and many women have joined the work force – often inadvertently neglecting children at home.

Also, in the same prosperous United States, the nation's poorest citizens are its youngest. "The 23% poverty rate for children under six is more than double the rate for adults. Despite the anti-poverty programs, and the general affluence of the country, the proportion of young children who live in poverty in the US is increasing." Is this a triumph of capitalism? Is this a path to emulate by others who embrace capitalism as the road to salvation?

Recent studies have shown that "America has virtually abandoned the seriously mentally ill who cannot afford to pay for treatment". More than 250,000 people with schizophrenia and manic-depressive psychosis live in public shelters, on the streets, and in prisons or jails. Many others among the United States' two million seriously ill, who cannot afford medical treatment, live in communities where no appropriate services are available. "The majority of all of these are receiving little or no psychiatric treatment, because most public psychiatric services have broken down completely." Is this another triumph of capitalism?

The ills of capitalism have grown. The poor and underprivileged are short-changed. Nothing trickles down. What has happened to the ideals of Liberty, Humanity and Fraternity under the banners of which capitalism was supposed to be acting? What has happened to the noble and inspiring notion: "Give me your tired, your poor, your huddled masses, yearning to breathe free," which was supposed to be the motto and symbol of the free and promising United States.

Are these but empty words? They were not meant to be empty words. The ideals of social justice were meant to be taken seriously. The idea that the poor and underprivileged are to be helped by those who are more fortunate was to be taken seriously. We are living in one web of life. It has been an accident of fortune that one person was born with a white face in an affluent suburb of Los Angeles, while another was born with a yellowish face in the midst of the poor of Calcutta. We owe compassion to each other.

We know that entrenched systems based on privilege and on the exploitation of others – however subtle and indirect this exploitation may be – are not likely to reform themselves, do not become compassionate by themselves. We need to help them, sometimes in a resolute and firm way. Here again, the relevance of our brother Francis is clear and unmistakable. We need a new Franciscan Revolution.

We need a strong social movement inspired by spiritual ideals to transform capitalism into a spiritual and reverential force. We need a new form of socialism. I am proposing *Franciscan Socialism* as a possible solution. The Marxist equation is simply too narrow. Material satisfaction and the abolition of private property do not automatically lead to the creation of a New Person, who would create a perfect social system in which no one is exploited.

The human equation is ultimately a spiritual one. Social justice must go hand in hand with the enhancement of human dignity. Our humanity ultimately resolves itself in our spirituality. Unless we conceive of a human being as a sacred particle in a sacred universe, the grounds of human dignity and ultimately of social justice will be thin and wanting. We need to create a reverential economics in order to avoid violent revolutions in the future.

Franciscan socialism is not a name for a conjuring act which tries to reconcile the contradictory elements, but an idea borne out of a deeper reflection on the plight of present Western civilization. Material progress is not sufficient as a vehicle to deliver us to the gates of paradise; it is not sufficient either in the traditional Marxist or capitalist systems. Consumption as a form of heaven is a travesty of genuine fulfilment.

We deserve a better fate than being manipulated by the brute dictates of faceless Communist bureaucracies, or manipulated by the subtle techniques of the advertising industry, each of which deprives us of genuine freedom of choice and of genuine spiritual fulfilment.

Our present battle is one for the meaning of life, and for regaining destiny over our lives and the lives of all species. We suffer the spiritual anomy, the existential anguish, the value vacuum. *Each is a result of the breakdown of a civilization which has separated itself from the rest of creation and which emptied itself of the spiritual content.* By reducing ourselves to consumptive mannequins, we inadvertently chose to inhabit the universe which is suitable for Skinnerian pigeons, but not for spiritually destined human beings.

What we witness nowadays is a triple crisis: the crisis of Christianity, the crisis of Marxism, and the crisis of capitalism. None of these religions (for each can be conceived as a religion) can take much solace from the collapse of any other. *We need, in fact, a new spiritual revolution.* We need a new form of spirituality which simultaneously ignites our imaginations, inspires our wills, and shows us the way to practical and lasting solutions of our most urgent problems.

That this spirituality can be inspired and powered by Franciscan ideals, there is no doubt. Yet, the spirituality which is uniquely relevant for our times cannot be just a resurrection of Franciscan teaching. We need a form of ecological spirituality which creatively resolves the agonizing problems of our battered earth and at the same time provides an existential guide to our individual lives. Spirituality is a crystallized essence of the human condition at a given time, as I have argued in Chapter 1. In our times the crusade for the preservation of the Earth is a spiritual project. Our present spirituality not only needs to be green at the edges but must be based on ecological consciousness. The healing of the planet and the healing of the self are at the centre of our spiritual quests. Ecological spirituality is the recognition of the fact that to serve God nowadays, we need to serve the Planet. It is in this sense that we are witnessing the greening of God.

Summary

To understand the way of God for Thomas Aquinas was to understand it intellectually, through the abstract categories of Aristotle's philosophy. In our times, God's way can best be understood through ecological reason which informs us about our responsibilities, about the necessity of compassion, stewardship, caring ... co-creating with God.

Ecology is a new form of awareness. To think ecologically is to think connectedly, holistically, reverentially. Ecology as reverential thinking imperceptibly merges with religious thinking. Religious thinking for our times must include the ecological dimension. Thus ecology and religion merge; that is when we realize that God's earth is threatened and that due respect for God, in our times, is tantamount to cleaning the earth and treating it reverentially.

Ecology and Christianity will be fused together: each is too important for our well-being to be left out. The image of the God of the Old Testament was born of concerns and problems different from ours. As our consciousness has changed – guided by new concerns and visions, of the interconnected and ecologically viable cosmos – so our image of God has changed, and thus the reality of God has changed. We need a new light to brighten up our present sombre horizons. The light of ecological spirituality, guided and inspired by the visions of simplicity of St Francis, offers us a promising new perspective as we enter the twenty-first century and indeed the Third Millennium.

As we are building a new millennium, we must not forget the poor, the forlorn, the downtrodden. The quest for justice is a part of God's design. The quest for universal justice is an inherent part of our spirituality. All great spiritual traditions seek and promote justice. If they do not, then their spirituality is in question. Even if we are unable to bring about justice and eradicate the existing injustices, we must seek justice. This is our responsibility, our honour, our calling; our mission. Deep down, Marx was a spiritual seeker. He really wanted to redress human injustice and bring about universal justice. But he didn't know how to do it. Hence his tragedy, and the tragedy of the twentieth century.

When we look at our religion, and especially at Christianity, from the perspective of the year 3000 CE, we see it profoundly changed. We are part of this change. We will find traditional religion, and especially Christianity, adaptable to these vast changes, if only because *we shall make Christianity so adaptable*. Our evolutionary mind (which is God-in-the-making) is extremely adaptable. It will make traditional religion adaptable.

We shall become favourite children of God by assuming more and more responsibility, and by exercising it. Through the transformation of our own consciousness, we shall live the new path, the spiritual Tao in which traditional symbols of Judaism and Christianity will be transformed into a new religious reality – evolving as our minds and reality do. We shall therefore be worshipping the creator by making more of his creature.

Every religion has various threads woven into it. At certain points of history, some of the threads, neglected and ignored in the patterns existing hitherto, begin to shine with a new beauty. And when people are wise, new patterns are woven around these shining threads. The teaching of St Francis is such a thread within Christianity – neglected and ignored for a while, but now shining with a new beauty. It is time that we wove a new pattern of Christianity around those shining Franciscan threads.

Francis is a patron saint of many religions. Rebuilding our religion around his ideals is honouring Gandhi and Hinduism, is honouring the Buddha and Buddhism, is honouring Mahevira and Jainism, is honouring the best of many other religions.

CHAPTER NINE

The Beautiful and the Sacred

From Plato to the Evolutionary Perspective

One of the most remarkable statements on beauty is that of Plotinus (203–70) who was in many ways Plato's follower. Plotinus wrote:

> The More beautiful a Thing is
> The More Intensively does it Exist

We want to give our consent to the statement and say 'Yes'. But why? One of the classical answers was that of Plato. For Plato beauty was God, or more precisely an embodiment of God, an embodiment of the Divine. So for Plato there is no problem: beauty is an aspect of the Divine. The sacred and the beautiful are aspects of each other. In response to Plotinus's statement, Plato would say, yes the beautiful is more real than non-beautiful because the beautiful is closer to the Godhead. For this reason its existence is more *real* than that of the object removed from God. We need to remember that in Plato's scheme the existence of the ideal forms, or simply Forms (of which Godhead is one), is more real than the existence of mere physical objects.

This answer is altogether too easy. We have lived with Plato, or under the shadow of Plato, for too long. We repeat old words and old schemata until they become a substitute for our own thinking. As a matter of fact Plato's answers are, by and large, the answers of traditional religions, be it Hinduism or Christianity. In each, the Divine Ground of Being (the Brahman and God) explains the nature of beauty. Beauty is an aspect of the divinity of God. God partakes in beautiful objects and because of that they are beautiful. The answer "God makes beautiful objects beautiful" is no longer satis-factory, especially in virtue of our new insights into the nature of evolution.

In this chapter we shall attempt to provide a new conception of beauty. I will argue that beauty and divinity come from the earth, from the *evolutionary* process. Plato has got it the wrong way about by imagining that objects become beautiful by some kind of radiation from Heaven, from above. The right perspective is to see how beauty and the sacred have emerged from below – out of the miasma of *evolutionary becoming*. Some of the conceptions presented here owe much to Teilhard, Vico and other evolutionary thinkers, especially Heraclitus. Yes, we are reclaiming the tradition of Heraclitus, which has been so profoundly eclipsed by Plato. Let us then build an evolutionary model of becoming within which the idea of beauty and of the sacred would make more sense than hitherto explained.

When we look at the ascent of life, we see the formless acquiring a form, we see the incoherent becoming coherent, we see the inarticulate becoming the articulate. The grandeur of life comes through the articulation of *structure*. Structure is the key to under-standing the evolutionary concept of beauty. From the structures of inorganic life, evolution builds structures endowed with life, then it builds the structures of beauty – through the process of continuous transcendence. (We are back to the arguments of Chapter 1.) Ultimately, structure becomes a ladder to heaven. At this point, life has acquired spiritual characteristics. It now presses forward to make itself more and more divine. Let us see how this process manifests itself in evolution.

The beginning of the articulation of structures is the sea endlessly rolling for aeons of times. Out of this process of rhythmic rolling came all the forms of life and of human architecture. The continuous waves of the oceans created the shell – on which the rhythm of the waves is so visibly imprinted.

The shell is the original geometry of the universe. In the beauty and exquisiteness of the shell – its serenity and symmetry – we witness the anticipation of future temples and other things of beauty.

It is a part of the same evolutionary rhythm – from the shell to our ribs, and then to the columns of the Parthenon. In creating the shell, evolution was already toying with the idea of the temple, and was already creating a rudimentary paradigm of beauty.

Temples are among the most significant structures evolved by human beings. They embody and express the great rhythms and symmetries of life. The secret of structures and their greatness lies in their *symmetry*. We find these symmetries fascinating and irresistible because deep down *it is evolution in us that responds to its evolutionary epic*. Thus we have the key to understanding beauty in evolutionary terms: the objects of beauty are those which are invariably based on intricate rhythms and symmetries. These rhythms and symmetries are life-enhancing. For this reason life in us responds to them so readily. We ourselves are part of this stupendous evolutionary epic, and our lives are based on and carried through endless cycles of rhythms and symmetries.

Even before the oceans became the cradle of the amphibian and mammalian life forms, the driving force of the evolution of the universe had chosen symmetry as its basic modus operandi. All scientific laws (as we have conceived of them) have fallen. That means that each bears exceptions and therefore none is absolute or ultimate. Except the law of symmetry. Only the law of symmetry holds universally: to every element there exists a contrasting element which holds the original element in balance. This is astonishing and awesome in its simplicity. Balance and harmony are based on symmetry. This is the root of our experience with beauty.

The primordial symmetry is then woven into life-enhancing rhythms, which form the basis of life-enhancing structures. The rhythm and symmetry may be altered, modified and imaginatively played with – as they are in great works of art and in life itself, particularly those forms of human life which are lived to the brim. In brief, *the beauty of man-made structures and cosmos-made structures has its explanation in the life-enhancing qualities of the rhythms and symmetries of the universe itself*. We do not need heavens above to explain the ascent of beauty. The evolutionary process itself is sufficient. Furthermore, it is inconceivable in our day and age to explain beauty outside the evolutionary process. Evolution is the maker of life, of things of beauty, of things Divine.

The radiance of life is the coherence and endurance of structures which are not only capable of withstanding various stresses of life but are also capable of an astounding variety of performances of which early biological life could not dream. When the range of these performances is so enlarged that the human structures blossom forth with religious significance (through temples and religious rituals), blossom forth through the objects of art, and of poetry (for poetry represents unique structures of the universe), and through other specific human structures, *then* human spirituality has arrived and beauty has become a vehicle of the sacred and the spiritual. *Beauty is but a collective term, summarizing all those processes through which structures have gradually acquired more coherence, endurance and capability – until they shine through with new qualities: the aesthetic, the religious, the artistic, the ethical.*

We have now arrived at a rational, evolutionary explanation of Plotinus's dilemma. *Why is it so that the more beautiful a thing is the more intensively it exists?* Because it possesses more evolutionary life in it; because it manifests the triumph of the evolutionary unfolding. Beauty represents evolutionary radiance – coherence, articulateness and performance born of the relentless zest of life to make more of itself. *Beauty is the result of an upward process of relentless transcendence.*

Thus, *a more beautiful object exists more intensively* (in comparison with a less beautiful one) *because it literally contains more life in it*; more life in the evolutionary sense, more life in the sense of radiance which the process of transcendence has brought about.

Our evolutionary perspective necessitates a reinterpretation of the whole Platonic tradition concerning beauty; as well as a reinterpretation of many religious traditions which explained beauty and divinity as coming from above, from heaven, from God. As we have said, Plato and all those who followed him, have got the whole process backwards. Plato was right to connect beauty with divinity. But he assumed that beauty comes *downwards* – from the Platonic heaven. In the evolutionary perspective beauty and divinity come from below, grow from the bowels of life, are the stuff of life, are not separate from it; are the acts of life blossoming with a new radiance. God is the radiance of life continually transcending itself. Thus, the divinity of life, and beauty accompanying it, emerges out of the roots of life. Beauty is not only an aesthetic description of objects but an *evolutionary* description, a recapitulation of the process of transcendence, an acknowledgement of this stage of life when it reaches spiritual qualities.

All that we have said applies to human lives. The more beautiful the life of a human the more intensively does it exist. It does so by incorporating the richness of the evolutionary life-flow. It does so by partaking in a variety of life-enhancing forces; by affirmation and celebration of life – not by abdication from life, or denial of life, or annihilation of it.

If we look from this perspective on our age, some conclusions inevitably thrust themselves upon us. Why the atrophy of meaning? Why the dwarfing of the phenomenon of man? *Because a crisis of beauty is a crisis of man; because the break up of beauty is a break up of the coherence of human life.* The loss of beauty – in our times – is tantamount to a loss of meaning, coming on the wheels of the loss of coherence. The loss of spirituality is one of the consequences. Coherence – meaning – beauty are the steps or the links of the chain.

Beauty is not a luxury but a necessity of human life. As oxygen is indispensable to our lungs so beauty is indispensable to the inner coherence of our lives. However, this is not how beauty is viewed in the second half of the twentieth century – even by the philosophers of art, the artists and art critics. Let us examine some deeper reasons for their collective blindness.

The Pathologies of the Present Scene

What we witness nowadays is not only our intellectual confusion about the meaning of art, but our soul's confusion about the meaning of life; not only a breaking up of the ideal of beauty but a breaking up of the coherence of our lives; not only the loss of traditional crafts, forced out by shoddy plastic goods, but the loss of spirituality and God. While we try to grasp the whole picture, let us not overlook the cleverness of the technological culture which has succeeded in redefining a multitude of things in its own image and often made us dupes of its limited and distorted perceptions.

This process of subtle but pervasive redefinitions had already started in the nineteenth century, if not earlier. I have shown how this process occurred with regard to intrinsic ethical values in my book *Eco-Philosophy, Designing New Tactics for Living*. Utilitarianism, and then relativism and nihilism take ascendence over intrinsic values which are gradually pushed to the margin – as insignificant and obsolete. In the twentieth century, this trend will lead to rampant hedonism on the one hand, and to the emergence of cost-benefit analysis, as a new ethics, on the other hand.

This process of annihilation of intrinsic values leads, in the second part of the twentieth century, to the devastation of aesthetic values and the corruption and mutilation of the idea of beauty. It is true that to be alive, art had to search for new forms of expression, new idioms, new renderings of reality and of the human experience. This search for new forms and new expressions however becomes rather desperate at the turn of the century and in the first decades of the twentieth century. Artists feel overwhelmed and suffocated by the weight of tradition, by the achievement and variety of traditional art. To break the creative paralysis, the avant-garde becomes bolder, more radical, more determined. Hence the experiments of Picasso, Bracque, Salvador Dali and Giacommetti in visual arts; of Berg, Webern and John Cage in music. Art frees itself from the control of traditional forms. Traditional aesthetic criteria are loosened and relaxed, including the criteria of beauty.

This much is comprehensible and part of the natural story of evolution. To be alive, art must renew its idiom. New experiments with art forms brought about new insight into the nature of human perception, new perspectives on human nature, new renderings of reality.

But then something happens in the 1960s. The avant-garde becomes a merry-go-round of consumption – of 'New Forms' invented every other year to provide the commercial market with 'New Art'. The meaning of the avant-garde is debased; the meaning of art cheapened; the phenomenon of man, as beneficiary of art, completely ignored.

While the avant-garde is breaking down under the assimilated strain of its meaningless experiments, another process is going on. In the 1920s an important artistic ideology emerges called 'the Bauhaus'. Later, it is called the International Movement. At first it is an architectural movement. Its motto is: *Form Follows Function*. This motto, after a while, becomes the chief aesthetic criterion. Traditional criteria of beauty are to be rejected. *Function* must be our guiding principle. Function itself must be rational and guided by the best technology available. Thus *the most advanced technological means begin to dictate how houses and man-made environments are designed and built*. In the 1960s, this whole process leads to the construction of soulless tower blocks, and sterile environments surrounding them. Technology has taken command, and is now trampling over our traditional artistic criteria, over our sense of beauty, over our sense of inner well-being. The results we all know from our own experience of the last quarter century.

The idea that form should follow function, that technology should determine our means and guide our praxis was not confined to architecture. It spilled over into other domains of human activities, including art. The criteria of excellence are more and more *dominated by technical excellence*. Artists bow to the technological imperative. Technology takes command more and more forcefully.

In this climate, traditional aesthetic criteria are further weakened, especially as art is being *used* in a new way – as a commodity, as an investment. In this new role art (and artists) are not being judged by their artistic excellence, by the intrinsic value of their art, by the life-enhancing quality of their pieces, but by the market value. The market value is invariably manipulated by gallery owners and money men, who often invest in art without the slightest interest in the intrinsic values of art.

Let us recapitulate. We have discussed two reasons for the plight of contemporary art and the perversion of the idea of beauty. The first reason is that the avant-garde has lost its sense of purpose and direction, meandering amidst meaningless experiments. The second important reason for the subversion of beauty is the fact that technology has become the guiding force of culture. Yet there is a third reason, more important than the other two. We would not have succumbed to the tyranny of the technological imperative, we would not have so easily tolerated the aberrations of the avant-garde, *if* the whole civilizational process had not become topsy-turvy and started to evolve along a pathological course.

During the last century we have witnessed a systematic process of dwarfing the entire phenomenon of man. The forces that have conspired in this process have been described in many books, including some of my own. Let it be noted that the destruction of intrinsic values in the nineteenth century and the withering of beauty and the loss of meaning in the twentieth century are aspects of the same process, are symptoms of the collapse of man as a transcendent and spiritual being.

I have argued elsewhere (in "Antinomies of Form in Contemporary Art") that form in art follows and reflects the human condition of a given time – just like spirituality does. The parallel between the two should not surprise us. On the contrary, it should surprise us if it were otherwise. When man became gradually diminished, form

in art and art itself began to reflect the condition of this impoverished creature called "technological man".

All the twists and convulsions of modern art may be said to be
an expression of life itself. But to say that much is to say too little.
*Wretched art expressing the wretched condition of man is not much
of a spectacle*. Present art cannot excuse itself for its shortcomings
by claiming that it "follows life", while life itself has degenerated
into a pathology. Let us see it clearly. The break–up of the coherence
of man, and an art expressing the agony of this break–up, are
parallel phenomena. They are parts of the same process – the
break–up of beauty. Because of the simultaneous break–up of the
coherence of man and the ideal of beauty, the conceptions of art
and its definitions become as twisted as the human condition itself,
become debased as human life itself, become polluted and uglified
as the condition of our being.

We have therefore arrived at a seemingly rational explanation for
the arbitrariness in the present use of the concept of beauty, including the apotheosis of ugliness as a manifestation of beauty. *Our
schizophrenic culture has lost its anchor, its direction, its purpose.
It has become violent and ugly. In the process, it has created art in
its own image.* Through its cleverness, our culture (all cultures are
clever, including technological culture) attempts to use us in order
to justify its ethos so that we become persuaded that ugly and
beautiful are interchangeable; and that the sublime and the brutal
are considered to be only matters of definition. In this way, the
soulless culture, which has destroyed our spirituality and broken up
the meaning of beauty, attempts to make us its willing and rational
accomplices. We are supposed to rationalize the pathology of the
culture, which expresses what "is", while deep down we feel present
art to be an abomination to our soul and a continuous offence to
our sense of beauty.

In summary, the technological society is not only plastic, superficial, slothful, irreverent. It is also devious and clever. It has
succeeded, or almost succeeded, in redefining culture or subduing
culture to its consumptive and power-crazy ethos. What we witness
during the latter part of the twentieth century is not so much the
desperation of avant-garde artists to express themselves in a new
way but a subtle subversion of art by the commercial world, a subtle
buy-out of artists to serve the overall goal of the technological
culture. When art glorifies the ugly and the violent, the violence and

ugliness of the outside world become aspects of the "Natural scene".

A death-ridden culture is a denial of beauty by necessity. Beauty is vibrancy. Beauty is life singing to the universe. Technological culture is none of that. Let us emphasize: beauty, meaning and spirituality are inherently connected as they spell out coherence, endurance, inner purpose, gazing at the stars and asking what's it all about. God makes coherence through the structures of beauty. Life makes the journey of transcendence through structures of increasing coherence and performance, and thus of beauty. Such has been the evolutionary story of life. Do we see the continuation of this story within the technological society? Hardly.

If this analysis of technology strikes some as a savage attack on technology, I can only respond that I have merely attempted to place a mirror in front of this unfeeling machine. Technology has been a savage god unsparing of human spirituality, of human tenderness, of everything that human cultures have deemed as great and noble.

The denial of beauty is the withering of spiritual life, is the undermining of human meaning. When beauty is rendered meaningless it loses its potency. Then the ugly is acceptable, then spiritual squalor and the meaninglessness of our lives becomes an "acceptable" consequence of a culture which has deliberately destroyed the concept of beauty.

Perhaps beauty had to be destroyed so that techno-culture could proceed unhampered in its destruction of natural environments and its uglification of our inner lives. The picture is now clear. It is clear, that is, if we have enough courage to look at the whole of civilization from a vantage point which sees technological culture for what it is.

It is silly merely to observe how the meaning of the term "beauty" has changed. We must be able to understand how the process of civilization has gone; and specifically why it has subverted and cheapened the human condition. Furthermore, we must have the courage to outline new roads leading to the recovery of human meaning, of human spirituality, of human beauty.

Beyond the Beautiful

What is divine, is beautiful. Why? What is God-like, is beautiful. Why? What is sacred, is beautiful. Why? The three questions are aspects of each other. There is a traditional answer to our questions.

God-like and divine are the embodiments of the highest perfection. What is absolutely perfect must also be beautiful.

Yet this answer is loaded, old-fashioned, biased from the start – in its language and in its conception of the world. It simply *assumes* that there is a God; secondly, that God is perfect; thirdly, that being perfect it is beautiful. Thus Divine is beautiful by its very definition. But this answer is too easy, as we have pointed out before. Besides, this answer does not answer our 'why' questions.

What is beautiful must be judged by *the criteria of human experience*. Human experience is the ultimate terminus point which decides upon the nature and meaning of beauty. We need to add, however, that what we have in mind is the experience of the whole race, *experience in the evolutionary sense*, not any whimsical, idiosyncratic, subjective experience.

When the early sages discovered the delight in contemplating the glory of the visible and invisible world, when their minds became sufficiently attuned to *name* things as beautiful and sacred – which became the highest attributes of human experience – then it was natural that they would extricate the inner experience of beauty and *project* it on the objects themselves. Thus objects were *endowed* with characteristics of beauty – as if they could be beautiful objects without the experience of beauty. Finally these early sages found the locus for the experience of Ultimate Beauty, this beauty which combines the aesthetically pleasing with awe and with total rapture. They projected this experience of Beauty on the special objects, which they started to call *deities*. After a time people began to imagine that these deities themselves were the source of the divinity and beauty which radiated from them on the human universe. Thus after a while they assumed that humans only receive beauty from above. We have just re-traced the God-centred explanation of the idea of beauty and the meaning of the sacred.

Seen from the evolutionary perspective, the whole scene looks different. It is human experience which is the ultimate maker of beauty and divinity. How does the sense of beauty (and the sense of the sacred) manifest itself within our experience? Each of us has experienced moments of beauty and also some moments of *sublime* beauty. In my life time, I have met at least three saints. They were humans whose state of being was so special that I would not hesitate to call this state of being Divine. These persons were the saint of Kanchipuram, in southern India, the patriarch of the Zen Buddhist temple at Eheiji in Japan, and Mother Teresa of Calcutta. Each completely different as human personalities. Yet united by the common thread – for which the best name is the divinity of being.

What was most striking in each of them was this special light which shone through them without any effort. If the universe came from light, and if God is light, they embodied this light to a much higher degree than I have found in any other human person.

The next characteristic I found in each of them was enormous simplicity. There is no other way of describing this form of behaviour which is so Divine and so natural at the same time.

Yet another striking trait in each of them was the sense of enormous freedom. They were completely liberated people, completely realized. Is it not how we think of God – as completely free and liberated? Is it not a defining aspect of divinity – this enormous freedom, unconstrained by any contingencies?

Yet another characteristic of their being was enormous peace. Living in a world of turmoil, in the second part of the twentieth century, one wondered how they could have been *so* peaceful. There was something God-like in their message of peace. Is it not how we imagine God, and Divinity in general – as unperturbed serenity?

Yet there was also the enormous courage expressed through their behaviour and their entire life story. Courage and simplicity combined; peace and freedom; and this continuous light emanating from them. All those aspects of their being, combined together, made them saints (divine beings) in my eyes. What else is divinity if not those extra-ordinary characteristics of those special exemplary individuals which we register and acknowledge within our own structure of experience? These characteristics we find so much out of the ordinary, so sublime that we are compelled to call them Divine or God-like. After we have experienced these sublime characteristics in other human beings, an abstract idea of Divinity is formed. *Divinity is a sum-total of the awesome characteristics of other human beings which we have experienced ourselves within our own scope of understanding.* All notions of Divinity bear the mark of human experience and human understanding. And it cannot be otherwise. What is Divinity for the bees, for the raccoons, or for extra-terrestrial intelligences, we have no idea.

This argument is not a repetition of Xenophanes's notion that people anthropomorphize their gods and make them in their own image. Xenophanes sarcastically pointed out that if horses invented a religion, their gods would be in the shape of the horses. Specifically, we are not suggesting or proposing that Divinity is a subjective product of the human mind. We rather suggest that the whole universe is Divine in nature, and Divinity or sacredness is spread throughout the cosmos and pervades it. But in order to receive the universe as Divine, however, we have to be sufficiently sensitive to these aspects of the universe which make it Divine. The universe has created us (and other beings) to celebrate its richness, and ultimately its Divinity. For Divinity, uncelebrated, unexperienced by any being, unrecognized by any intelligence, is Divinity in *status nascendi*.

I do not believe that the bullocks and cows, which were quietly chewing their pasture in their sheds within the walls of the Mandir, where the Saint of Kanchipuram was blessing us with his presence, could sense or experience the Divinity of the saint. The cow cannot experience Divinity in the sense we use the term. God and the universe need us to manifest their Divinity for we make this Divinity explicit. *The unfolding of evolution means the actualization (through us) of the Divine which is potentially contained in the cosmos.*

Let us trace the manifestation of the Divine in another sphere – in music. Wolfgang Amadeus Mozart. The sound of the name itself rings with celestial notes. If any music deserves to be called Divine, it is Mozart's. But *how* is this Divinity expressed? Has God descended from heaven to dwell in Mozart's notes? Blessed Mozart to have God write his music for him ... Or have Mozart's notes been so exquisitely rendered, so divinely shaped out of the stubborn melodic material that we hear in them God singing? Indeed, we hear in them our inner God crying in ecstasy.

Sounds into bars. Bars into melodic lines. Melodic lines into contrapuntal structures. *At what point* do these structures become the structures of beauty? *At what point* are these structures of beauty touching the Divine, or are touched by the Divine, or are an expression of the Divine? Here we approach the core of the human mystery and the mystery of the Divine.

Many around Mozart were astonished and envious. How can a *human* being write such heavenly music? Beethoven was not disturbed. He went on to participate in the Divine in his own way. The mystery for us is: did Mozart and Beethoven *participate* in the Divine? Or did they *create* the Divine? Or were they simply vessels for the Divine to pour into? To choose the third option is altogether too easy. It resolves the problem by explaining nothing.

Our language at this stage of our intellectual journey is still insufficient. It cannot adequately express the subtlety and power of the participatory process, whereby evolutionary structures, by ever-transcending themselves, begin to sing with the music of Mozart, begin to enrapture us with the hymn to brotherhood of Beethoven's *9th Symphony*. Mozart's *Requiem* and Beethoven's *9th* are the evolutionary structures through which life sings with beauty and Divinity. The genius of Mozart and Beethoven lies in their supreme understanding of the subtlety and power of human-made structures.

These structures are the expression of cosmic energy – because there is no other. They are so *exquisitely woven* that we are compelled to call them Divine. *It is there, in the weaving itself, that Divinity lies.* For what is this weaving if not an evolutionary process, which continuously transcends itself through ever more subtle, coherent, enduring, beautiful structures?

The cosmos reverberates with divine energies. Mozart and Beethoven captured this energy in some of their works. Saints have translated this energy into their life-forms. To partake in the Divine, your consciousness must be so finely tuned that it is divine itself. The great mystery of the evolutionary process is that it can produce consciousness which is capable of embracing the divine, by being itself an aspect. *The closer we come to the Divine the more we can experience the Divinity of the Universe.*

We have thus closed the circle. The sacred cannot be but beautiful because whenever it manifests its presence to us, it uses the structure of beauty as its vehicle. But let us be clearly aware of the nature of the argument. We are not saying that beautiful is beautiful because the sacred (or the Divine) has found a dwelling in it. We are saying – in accordance with our evolutionary perspective – that in their evolutionary ascent, these structures which finally reached divinity are the ones which, on their way, passed through the stage of beauty. Thus, beauty is a vehicle of the sacred, an inherent aspect of it. The sacred embodies the beautiful and transcends it at the same time. What is this aspect of the sacred that is beyond the beautiful is precisely the quintessence of the sacred, and is precisely this aspect of the sacred that is beyond words.

The evolutionary conception of divinity explains the maturing of divinity throughout the evolution of life: from dim fragments of consciousness, life proceeds to full consciousness, to aesthetic consciousness, to ecological consciousness, to divine consciousness. With the help of divine consciousness, we make the world divine. When we possess divine consciousness, we are then divine.

In-between the brute atoms and God, there is the stage at which we are at present – struggling with our own identity, with our future destiny, with our religion, with our God, with our inner God, with our spirituality. We have nowhere to go but upwards. In this journey of self-realization and God-making, we include the Earth – our outer home. The sacred Earth is the realm between the sacredness of our soul and the sanctity of the whole universe.

Summary

As life develops, it creates more coherent and enduring structures. As these structures become more and more versatile, they blossom forth as symbolic structures of art, of religion, of human spirituality. The structures of great temples epitomize the spiritual ascent of evolution. The physical has been subtly transformed into the symbolic and spiritual.

> Seen in the evolutionary ascent
> Beauty is the structure
> of coherence
> of endurance
> of radiance ...

When the radiance becomes overwhelming it becomes divinity. Divinity is the blossoming forth of the forms of transcendent beauty which left behind the physical to become spiritual.

The age-old dilemma – why more beautiful objects exist more intensively than less beautiful ones – is now resolved. Because there is more evolutionary life in them. Beauty means life. The more intensive a beauty the more layers of life it contains. When beauty atrophies, human life atrophies. A crisis of beauty is a crisis of man. Beauty is not a luxury but a necessity of human life. A culture which has lost its sense of beauty, has lost its sense of purpose, has lost its sense of spirituality. Beauty does not exhaust the sense of the spiritual, but is an important aspect of it. Spirituality and beauty are hymns to the Divine, are parts of the music of the Divine. The Divine manifests itself through the structure of beauty. Beauty is an outward appearance of the Divine. But this appearance is not accidental. It is essential to the existence of the Divine – as long as it manifests itself through the human mind and soul.

CHAPTER TEN

The New Gospel

The New Gospel

The World is a Sanctuary. The world is not a machine but an exquisite sanctuary. To treat yourself well, to treat others well, you must know that the world is not a heap of meaningless rubbish but a place reverberating with divine energies. The world is a sanctuary.

You need to be one with the world. You need to feel that you are not an accident in this world. You need to feel that the world is not an accident either. You need to celebrate your own uniqueness in the world that is sacred. The sacredness of the world, your dignity and your sovereignty, are assured by the assumption that the world is a sanctuary.

The universe conceived as a sanctuary gives you the comfort of knowing that you live in a caring, spiritual place, that the universe has meaning and your life has meaning. *To act in the world as if it were a sanctuary is to make it reverential and sacred*; and it is to make yourself elevated and meaningful. What the universe becomes depends on you. Treat it like a machine and it becomes a machine. Treat it like a divine place and it becomes a divine place. Treat it indifferently and ruthlessly and it becomes an indifferent and ruthless place. Treat it with love and care and it becomes a loving and caring place.

You were born creative. This is part of your likeness to God. Your creative nature is your gate to freedom. Be more free by consuming less. It is not right that you should be a slave manipulated by others who want you to be less critical and less creative, so that you consume more. As the spider reaches the liberty of space by means of its own thread so the thinking person reaches freedom by means of the renunciation of unnecessary consumption.

You hold Destiny in your hands. Not only your own but the destiny of those with whom you are interlinked. You hold the future of the planet in your hands – the beautiful Gaia, to which you owe so much. You hold God's entire creation in your hands. We are not only God's children. We are his messengers, his shepherds, his co-workers. You co-create with God. This is why God gave you intelligence. This is why God made you creative. You safeguard God's law by not letting others spoil or destroy God's beautiful planet. Be truly creative by understanding that God wants you to co-create with *him*, that God wants you to protect creation which is the source of all creative powers.

You have the responsibility to do your part. Yes, you personally. You cannot hide forever by blaming the system, the industry, the bad people, the bigness of it all. You are an intelligent person in this world. This world will be renewed by intelligent people like you. You have always known that you have the responsibility. Now is the time to exercise it. You have always known that consumption is a trap and does not make you truly happy. You have always known that you have the spiritual potential within you which may be called upon one day. Now is the time to release this spiritual potential. Now is the time to fulfil your obligations to yourself, to others, to God. Yes, the greater the responsibility you assume the more God is delighted with you. If you do nothing or close yourself in the shell of your egotism, you are of no use to anybody, you are of no use to God. To be truly happy you must accept your share of responsibility. The greater the share of responsibility you take, the greater person you become, the closer you come to God. Look at Mother Teresa, look at Gandhi – such simple people. Such magnificent people. So blessed by God. Why? Because they embraced the responsibility for all. They burned with their responsibility like a torch. But not in self-sacrifice – in the deepest self-fulfilment. To the glowing delight of God.

The web of life includes all forms of life, human and non-human. St Francis speaks of our sister river, our brother fire and Sir Brother Sun. All living beings, the foxes, the ravens, the wolves and the deer were included in his family. And so they should be included in our family. *Know thy place in the web of life.* It is your nourisher, your sustainer, your family. You are surrounded by countless brothers and sisters in creation on whom you depend so vitally. Life is all-inclusive. And so our spirituality and our religion must be. It must include all forms of spiritual life. It must be tolerant to different images of God. Different spiritualities and different images of God are all branches of the divine tree. Let us forgo what divides us in our spiritual traditions. Let us cultivate what unites us. What unites us is the love of life, the love of this beautiful planet, the love of the magnificent universe which has begotten the planet and then us as a part of its divine flowering. Ecology is what unites us all. Ecological faith is what expresses our deepest bond to the planet Earth and to God who created it all. Let us not worship new gods. But let us worship the true God who is one of compassion, justice, of ecological piety as *he* wants us to help him to reconstruct and heal the planet.

Be compassionate to others. Consume less so that others can simply survive. We now know that we live in one interconnected net. What we do in our corner of the net vitally affects what happens in other regions of the net. If we consume too much, use too much energy, produce too much waste, this will affect other weaker people, in other parts of the globe, who will be forced to sell their raw materials and to whom we will export our waste. This waste will return to us later. It will find its way to the food chain – and we will eat our poison – as we import food from other people.

We cannot live in peace on this planet until we have justice – justice not only for a select few, but for all people and for all creation. In the name of justice – human justice and God's justice – be frugal and compassionate. Help the planet, help other people by demanding less, consuming less – while you become a more spiritual person. The more you consume the less spiritually attuned you become. The more spiritually attuned you are, the more mindful, the more compassionate, the more frugal you become; and therefore more just. Frugality and justice are firmly connected nowadays. God calls you to render his justice by being radiantly frugal. Frugality is grace without waste.

Be gentle to yourself. You have endured enough stresses and unnecessary battering. You are tough. You can take a lot. But by taking too much, your heart hardens too much. You must have a gentle heart so that you behold yourself and others compassionately. Holding destiny in your hands means beholding yourself and others compassionately.

Be mindful how you treat your body. Don't abuse it. It is God's gift. Your frame is divine. Treat it as such. If you overeat constantly, it is not good for your health. You are then careless about your divine gift. Besides, the food that you overconsume is probably taken from the mouths of those who will go to sleep hungry tonight. Golden medium is our path. This is what great spiritual traditions recommend: neither excessive austerities nor excessive indulgence. On the wall of one of the Franciscan monasteries we read: "What you have and you don't need is stolen from those who need it and don't have it." Does this sound too radical for you? Then think again and think deeper. Maybe you will hear the ring of truth.

Be mindful of what you think and what you eat. If there is too much rubbish in your head, then there is no room for God in it. It is said in a sacred scripture: "If men thought of God as much as they think of the world, who would not attain liberation?" That is one of the cosmic laws. Keep the mind pure, for what a person thinks, he becomes. How much time do you allow yourself to dwell in the space of God? You claim to have no time. You claim that there is too much pressure on you. But these are the excuses for not attending to what is vitally important to you. You *can* find time to talk to your inner self. And then God will speak to you. Those who do not have time for their inner selves, God ignores. Speak more often to God and less often to the television. You will then find yourself in the space of peace.

You were born into a beautiful world. Yes, the world is scarred, but still beautiful. The morning dew on the blades of grass in early June, the smell of hay after the grass has been cut and is contentedly drying, the darkness of a big forest confiding to you its mysteries, the joyous clouds playing in the high stratosphere to make you light and joyous are all signs of the beauty of the visible world.

We have *spoilt* a great deal of this beautiful world. The rivers that were once so clean that you could not only swim in them but also drink water from them have become dirty pools. But not all is lost. We can renew what we have spoilt. We can bring back radiance to the earth. We can clean polluted skies, polluted lands, our polluted minds. We can do it if we remember what an incredible creation the earth has been; what an incredible creation we ourselves are! We have the will and the energy to do it – if we have the vision. Let us put our wills together to bring back radiance to this incredible world so that the earth is good and fair again.

Your nature is divine. The whole visible world around you is divine. The miracle of a magnificent oak tree growing out of a humble acorn is incredible. The whole universe is divine in its forms. You stand in front of an unending glory of continuous creation in which all the divine forms sing and dance to enchant and delight you. Remove the scales from your eyes so that you can see.

You are more aware of your divinity than the crab or the oak tree. Hence your responsibility is so much greater. Celebrate your divinity on behalf of those forms which are less conscious. How do you celebrate? By being responsible, by being reverential, by beholding yourself as a sacred particle in the sacred universe. When you are awakened from the stupor of unseeing, you are bound to embrace this divine cosmos with joy. For this divine cosmos is meant for joy and celebration. We do not live in a valley of misery and tears. Celebrate by renewing the Earth, by renewing yourself, by paying homage to God – while co-creating with him.

Your divinity must reveal itself in your action. Allow the inner voice to speak. And follow it. Link yourself with the highest light and become it. Don't be ashamed of your dignity and greatness. Being human you are great. Being divine you are great. Don't allow yourself to be cheapened or trivialized by consumerism and ordinary stupidities. Being trivial or cheap is not your destiny. This is not what you want to be. This is not God's plan for you. God's plan and Your Inner Voice are one. Listen to both attentively. And don't be afraid of being human – thus divine. You need to fulfil your inner mission, your divine potential.

Suffering cannot be avoided. Suffering should be diminished and reduced as far as possible. But suffering cannot be avoided. It is part of our human world – just as joy is part of our human world. And there is a time for each.

Through suffering you learn your humanity. Through suffering you learn compassion which enables you to understand the suffering of others. Through suffering you learn wisdom. You cannot live meaningfully in this world without some wisdom. You cannot acquire wisdom without some suffering. It is thus that we should consider suffering – as a teacher of wisdom and compassion, not of spurious pain. Those deep moments of solitude and suffering are necessary to make you a deeper person. Thus suffering is a cauldron in which your humanity acquires shape and makes itself divine. Those who have suffered a lot, understand a lot. They are the special children of God who went through many trials because they needed to learn much.

Would it not be better if from the start God had eliminated suffering altogether? This is not how this world works. It works through becoming. All forms of life mature through becoming. Going through the process of becoming includes suffering. For every becoming entails some pain. When the original acorn, which is about to become a mighty oak, splits up to give rise to a new tree, it suffers the pain of birth, the pain of becoming. Suffering and becoming, maturing and acquiring wisdom are part of the same process of life. You cannot avoid the life process while you are alive. Thus you cannot avoid suffering. Embrace it gently, but without being too fond of it. Ultimately, suffering must be transcended in favour of the glowing radiance of life – of which we all partake.

The fact of death cannot be avoided. Lead your life in such a way that you can meet death with serenity and dignity. What is on the other side of the curtain is a mystery. Only on the other side will you know why you couldn't know what is there from this side.

You should not fear death. Your entire life should be a training for death. Leading life as a preparation for death makes your daily chores so much easier. You then have a perspective – how to look at things which are important and which are not. Remember, at one point you will need to leave behind all your possessions and all your earthly ties. You must be serene about this. Such is life. It continuously metamorphosizes itself in endless cycles. You belong to one cycle. But your energy will be circulated in the universe forever.

Celebrate! The universe is in a state of self-celebration. The universe is a place of suffering. Thus we were born to suffer. The universe is a place of celebration. Thus we were born to celebrate. Hence the inevitability of suffering and the inevitability of celebration. The Sufis, the Christian mystics and other ecstatics know this truth well, namely, that the universe is for celebration!

It is easy to become overwhelmed by suffering. It is natural to recognize the suffering of others – as the bond that unites us. But there also exists another bond – the bond of ecstatic celebration which unites all the pulsating cosmos. Every living cell in the human organism, every bird in the sky and every tree in a healthy forest is singing the song of celebration. The greatness of the creator reveals itself in the fact that it created the universe which realizes itself through celebration. To acknowledge that the universe is in a state of perpetual celebration is to acknowledge its unspeakable grandeur, is to acknowledge the creative powers of God. True religion must be one which expresses and embodies this celebration – as we co-create with God and with the universe.

We are the glowing particles of the radiant universe. We are the cosmic dancers expressing the universe's ecstatic energy. We are the poets through whom the universe expresses its poetry. Through us the universe and God are singing their cantatas and fugues. The universe is a place of suffering. The universe is a place of celebration. One does not negate the other. Yet through celebration we redeem our suffering.

What is your path of liberation? To begin with, you need to *take yourself seriously*. What does it mean to take yourself seriously? Certainly not to be a pompous ass, full of gas and self-importance. But equally certainly, not to become a cipher manipulated by others. To take yourself seriously is to attempt to realize your potential. Yes, intellectual potential. But also your *ultimate* potential. To realize your ultimate potential is to become *a realized person and an enlightened being*. You can never recapture your divinity if you never assume that you are divine.

To take yourself seriously is to aspire to become God – without hallucinations, without megalomania, without losing your balance; by following a sane path, which becomes a path of grace and ultimately the path of divinity. Yes, in your daily life you will find that you so often depart from the path by being knocked off by external circumstance or your own indolence. The point is not to mortify yourself by crying, "Oh, I have failed again!" The point is to quietly return to the path. Centre yourself. And keep going.

Be aware of the awesome beauty of the universe – as it unfolds through you. We are the divinity unfolding. There is hope. And there is future – if we take ourselves seriously. We shall straighten up what is crooked. We shall purify and heal what is stained and polluted – if we take ourselves seriously, if we follow our divinity.

When can you hope to reach enlightenment? That depends how seriously you take yourself. In the meantime, you have the teaching of the Buddha, of Jesus, of Gandhi, of Krishnamurti, and yes, of Eco-philosophy.

Oikos – a Sacred Enclosure. Oikos is the Greek term for "home". We have to make our Oikos sacred. As the universe evolves, it acquires new qualities, which finally make our Oikos sacred. *Theos then is the highest quality of our existence as lived in the Oikos*: our home, nature, ecological habitats, the cosmos surrounding us.

When our home becomes a sacred enclosure, *Oikos* and *Theos* are united, are fused together, and become aspects of each other. Oikos is both a dwelling and a temple. Our responsibility is to take care of our Oikos, both in the immediate boundaries – of our home, and in its outer boundaries – of the universe which has nurtured us.

Achieve wholeness through your own effort. Unless we achieve wholeness through our own effort, no one can bestow this wholeness on us. We are living in the universe of Karma – as you sow so you will reap; not in the universe of revelation and miraculous salvation through the effort of others. Wholeness is reaching out to the transcendental heaven, to the sacred, while your feet are firmly on the ground, and your own person is in the service of others. Wholeness is this peace within, which connects you with the sacred but at the same time attunes you perfectly to your own body. Thus the existential and the theological become one.

We are the meaning makers. We ascribe meaning to the cosmos. In this sense we are cosmos makers. Because we are meaning makers, we are cosmos makers. Because we are cosmos makers we are meaning makers. We are neither God nor beast but only the median in between. By creating fragments of Grace within ourselves, we move from beasts towards the other end. Not to create these fragments of Grace is to betray our Divinity, our Humanity.

From the abyss of nothingness we have emerged. It is not our destiny to return there. We must create new forms of light. For we are light transforming itself. We are the photon transformed into the segments of grace. As our condition changes from the amoeba to sentient beings, we enlarge our consciousness. As we enlarge our consciousness, we reach upwards towards self-consciousness. As we acquire self-consciousness, we reach for understanding. As we reach for understanding, we reach for meaning. As we reach for meaning, we reach for grace. As we reach for grace, we reach for God. All is natural, and in a sense inevitable – when evolution creates the mind that becomes a source of light.

A PRAYER

The Divine Cosmos be kind to us.
Give us pure air, pure water, pure soil
So that we renew ourselves to serve thee.
Give us the light of understanding
So that we see our place in the Web of Life.
Give us humbleness of heart and the fire of vision
So that we work on thy behalf intelligently.
Give us our daily bread but also our daily inspiration
So that we do not sink into the consumerist slumber.
We ourselves promise to obey the Divine Law
Which is to cooperate, to serve, to share.
We promise to be the true children of God
By safeguarding God's universe.
We promise to be mindful of our daily steps
So that we do not trample over those weaker than we are.
We promise to behave compassionately and lovingly
For the sake of all creation, and our own sake.
To live in peace among those who hate
Oh, what a joy!
To live in compassion among those possessed by greed and avarice
Oh, what a joy!
To live with good values among those afflicted by cynicism
Oh, what a joy!

Index